Reviews of Environmental Contamination and Toxicology

VOLUME 246

More information about this series at http://www.springer.com/series/398

Reviews of Environmental Contamination and Toxicology

Editor
Pim de Voogt

Volume 246

 Springer

Coordinating Board of Editors

ISSN 0179-5953   ISSN 2197-6554   (electronic)
Reviews of Environmental Contamination and Toxicology
ISBN 978-3-030-07398-5   ISBN 978-3-319-97740-9   (eBook)
https://doi.org/10.1007/978-3-319-97740-9

Foreword

International concern in scientific, industrial, and governmental communities over traces of xenobiotics in foods and in both abiotic and biotic environments has justified the present triumvirate of specialized publications in this field: comprehensive reviews, rapidly published research papers and progress reports, and archival documentations These three international publications are integrated and scheduled to provide the coherency essential for nonduplicative and current progress in a field as dynamic and complex as environmental contamination and toxicology. This series is reserved exclusively for the diversified literature on "toxic" chemicals in our food, our feeds, our homes, recreational and working surroundings, our domestic animals, our wildlife, and ourselves. Tremendous efforts worldwide have been mobilized to evaluate the nature, presence, magnitude, fate, and toxicology of the chemicals loosed upon the Earth. Among the sequelae of this broad new emphasis is an undeniable need for an articulated set of authoritative publications, where one can find the latest important world literature produced by these emerging areas of science together with documentation of pertinent ancillary legislation.

Research directors and legislative or administrative advisers do not have the time to scan the escalating number of technical publications that may contain articles important to current responsibility. Rather, these individuals need the background provided by detailed reviews and the assurance that the latest information is made available to them, all with minimal literature searching. Similarly, the scientist assigned or attracted to a new problem is required to glean all literature pertinent to the task, to publish new developments or important new experimental details quickly, to inform others of findings that might alter their own efforts, and eventually to publish all his/her supporting data and conclusions for archival purposes.

In the fields of environmental contamination and toxicology, the sum of these concerns and responsibilities is decisively addressed by the uniform, encompassing, and timely publication format of the Springer triumvirate:

Reviews of Environmental Contamination and Toxicology [Vol. 1 through 97 (1962–1986) as Residue Reviews] for detailed review articles concerned with any aspects of chemical contaminants, including pesticides, in the total environment with toxicological considerations and consequences.

Bulletin of Environmental Contamination and Toxicology (Vol. 1 in 1966) for rapid publication of short reports of significant advances and discoveries in the fields of air, soil, water, and food contamination and pollution as well as methodology and other disciplines concerned with the introduction, presence, and effects of toxicants in the total environment.

Archives of Environmental Contamination and Toxicology (Vol. 1 in 1973) for important complete articles emphasizing and describing original experimental or theoretical research work pertaining to the scientific aspects of chemical contaminants in the environment.

The individual editors of these three publications comprise the joint Coordinating Board of Editors with referral within the board of manuscripts submitted to one publication but deemed by major emphasis or length more suitable for one of the others.

Coordinating Board of Editors

Preface

The role of *Reviews* is to publish detailed scientific review articles on all aspects of environmental contamination and associated (eco)toxicological consequences. Such articles facilitate the often complex task of accessing and interpreting cogent scientific data within the confines of one or more closely related research fields.

In the 50+ years since *Reviews of Environmental Contamination and Toxicology* (formerly *Residue Reviews)* was first published, the number, scope, and complexity of environmental pollution incidents have grown unabated. During this entire period, the emphasis has been on publishing articles that address the presence and toxicity of environmental contaminants. New research is published each year on a myriad of environmental pollution issues facing people worldwide. This fact, and the routine discovery and reporting of emerging contaminants and new environmental contamination cases, creates an increasingly important function for *Reviews.* The staggering volume of scientific literature demands remedy by which data can be synthesized and made available to readers in an abridged form. *Reviews* addresses this need and provides detailed reviews worldwide to key scientists and science or policy administrators, whether employed by government, universities, nongovernmental organizations, or the private sector.

There is a panoply of environmental issues and concerns on which many scientists have focused their research in past years. The scope of this list is quite broad, encompassing environmental events globally that affect marine and terrestrial ecosystems; biotic and abiotic environments; impacts on plants, humans, and wildlife; and pollutants, both chemical and radioactive; as well as the ravages of environmental disease in virtually all environmental media (soil, water, air). New or enhanced safety and environmental concerns have emerged in the last decade to be added to incidents covered by the media, studied by scientists, and addressed by governmental and private institutions. Among these are events so striking that they are creating a paradigm shift. Two in particular are at the center of ever increasing media as well as scientific attention: bioterrorism and global warming. Unfortunately, these very worrisome issues are now superimposed on the already extensive list of ongoing environmental challenges.

The ultimate role of publishing scientific environmental research is to enhance understanding of the environment in ways that allow the public to be better informed or, in other words, to enable the public to have access to sufficient information. Because the public gets most of its information on science and technology from internet, TV news, and reports, the role for scientists as interpreters and brokers of scientific information to the public will grow rather than diminish. Environmentalism is an important global political force, resulting in the emergence of multinational consortia to control pollution and the evolution of the environmental ethic. Will the new politics of the twenty-first century involve a consortium of technologists and environmentalists, or a progressive confrontation? These matters are of genuine concern to governmental agencies and legislative bodies around the world.

For those who make the decisions about how our planet is managed, there is an ongoing need for continual surveillance and intelligent controls to avoid endangering the environment, public health, and wildlife. Ensuring safety-in-use of the many chemicals involved in our highly industrialized culture is a dynamic challenge, because the old, established materials are continually being displaced by newly developed molecules more acceptable to federal and state regulatory agencies, public health officials, and environmentalists. New legislation that will deal in an appropriate manner with this challenge is currently in the making or has been implemented recently, such as the REACH legislation in Europe. These regulations demand scientifically sound and documented dossiers on new chemicals.

Reviews publishes synoptic articles designed to treat the presence, fate, and, if possible, the safety of xenobiotics in any segment of the environment. These reviews can be either general or specific, but properly lie in the domains of analytical chemistry and its methodology, biochemistry, human and animal medicine, legislation, pharmacology, physiology, (eco)toxicology, and regulation. Certain affairs in food technology concerned specifically with pesticide and other food-additive problems may also be appropriate.

Because manuscripts are published in the order in which they are received in final form, it may seem that some important aspects have been neglected at times. However, these apparent omissions are recognized, and pertinent manuscripts are likely in preparation or planned. The field is so very large and the interests in it are so varied that the editor and the editorial board earnestly solicit authors and suggestions of underrepresented topics to make this international book series yet more useful and worthwhile.

Justification for the preparation of any review for this book series is that it deals with some aspect of the many real problems arising from the presence of anthropogenic chemicals in our surroundings. Thus, manuscripts may encompass case studies from any country. Additionally, chemical contamination in any manner of air, water, soil, or plant or animal life is within these objectives and their scope.

Manuscripts are often contributed by invitation. However, nominations for new topics or topics in areas that are rapidly advancing are welcome. Preliminary communication with the Editor-in-Chief is recommended before volunteered review manuscripts are submitted. *Reviews* is registered in WebofScience™.

Inclusion in the Science Citation Index serves to encourage scientists in academia to contribute to the series. The impact factor in recent years has increased from 2.5 in 2009 to 7.0 in 2017. The Editor-in-Chief and the Editorial Board strive for a further increase of the journal impact factor by actively inviting authors to submit manuscripts.

Amsterdam, The Netherlands   Pim de Voogt
August 2018

Contents

Contributors

Hans Allmendinger Currenta GmbH & Co. OHG, Leverkusen, Germany

Mark Annevelink KWR Watercycle Research Institute, Nieuwegein, The Netherlands

Department of Environmental Science, Radboud University Nijmegen, Nijmegen, The Netherlands

Christian Boegi BASF SE, FEP/PA - Z570, Ludwigshafen, Germany

Bart T. A. Bossuyt Huntsman Europe, Everberg, Belgium

Wan-Loy Chu School of Postgraduate Studies, International Medical University, Kuala Lumpur, Malaysia

Kelsey Craig California Department of Pesticide Regulation, Environmental Monitoring Branch: Air Program, Sacramento, CA, USA

Ann-Hélène Faber Copernicus Institute of Sustainable Development, Faculty of Geosciences, Utrecht University, Utrecht, The Netherlands

KWR Watercycle Research Institute, Nieuwegein, The Netherlands

Institute for Biodiversity and Ecosystem Dynamics, University of Amsterdam, Amsterdam, The Netherlands

Herman Kasper Gilissen Utrecht Centre for Water, Oceans and Sustainability Law, Faculty of Law, Economics and Governance, Utrecht University, Utrecht, The Netherlands

Bjoern Hidding BASF SE, RB/TC - Z570, Ludwigshafen, Germany

Yih-Yih Kok Applied Biomedical Science and Biotechnology Division, School of Health Sciences, International Medical University, Kuala Lumpur, Malaysia

Michiel H.S. Kraak Institute for Biodiversity and Ecosystem Dynamics, University of Amsterdam, Amsterdam, The Netherlands

Choy-Sin Lee Department of Pharmaceutical Chemistry, School of Pharmacy, International Medical University, Kuala Lumpur, Malaysia

Rosanne J. Michielsen Institute for Biodiversity and Ecosystem Dynamics, University of Amsterdam, Amsterdam, The Netherlands

John R. Parsons Institute for Biodiversity and Ecosystem Dynamics, University of Amsterdam, Amsterdam, The Netherlands

Marleen van Rijswick Utrecht Centre for Water, Oceans and Sustainability Law, Faculty of Law, Economics and Governance, Utrecht University, Utrecht, The Netherlands

Paul Schot Copernicus Institute of Sustainable Development, Faculty of Geosciences, Utrecht University, Utrecht, The Netherlands

Thomas Schupp Faculty of Chemical Engineering, Muenster University of Applied Science, Steinfurt, Germany

Judy Shamoun-Baranes Institute for Biodiversity and Ecosystem Dynamics, University of Amsterdam, Amsterdam, The Netherlands

Summer Shen Dow Chemical (China) Investment Limited Company, Shanghai, China

Bernard Tury (Former) International Isocyanate Institute Inc., Boonton, NJ, USA

Pim de Voogt KWR Watercycle Research Institute, Nieuwegein, The Netherlands

Institute for Biodiversity and Ecosystem Dynamics, University of Amsterdam, Amsterdam, The Netherlands

Jun-Kit Wan School of Postgraduate Studies, International Medical University, Kuala Lumpur, Malaysia

Robert J. West International Isocyanate Institute Inc., Boonton, NJ, USA

Annemarie van Wezel Copernicus Institute of Sustainable Development, Faculty of Geosciences, Utrecht University, Utrecht, The Netherlands

KWR Watercycle Research Institute, Nieuwegein, The Netherlands

How to Adapt Chemical Risk Assessment for Unconventional Hydrocarbon Extraction Related to the Water System

Ann-Hélène Faber, Mark Annevelink, Herman Kasper Gilissen, Paul Schot, Marleen van Rijswick, Pim de Voogt, and Annemarie van Wezel

Electronic supplementary material The online version of this article (https://doi.org/10.1007/398_2017_10) contains supplementary material, which is available to authorized users.

Ann-Hélène Faber (✉)
Copernicus Institute of Sustainable Development, Faculty of Geosciences, Utrecht University, Utrecht, The Netherlands

KWR Watercycle Research Institute, Nieuwegein, The Netherlands

Institute for Biodiversity and Ecosystem Dynamics, University of Amsterdam, Amsterdam, The Netherlands
e-mail: a.faber1@uu.nl

M. Annevelink
KWR Watercycle Research Institute, Nieuwegein, The Netherlands

Department of Environmental Science, Radboud University Nijmegen, Nijmegen, The Netherlands
e-mail: Mark.annevelink@kwrwater.nl

H.K. Gilissen · M. van Rijswick
Utrecht Centre for Water, Oceans and Sustainability Law, Faculty of Law, Economics and Governance, Utrecht University, Utrecht, The Netherlands
e-mail: h.k.gilissen@uu.nl; h.vanrijswick@uu.nl

P. Schot
Copernicus Institute of Sustainable Development, Faculty of Geosciences, Utrecht University, Utrecht, The Netherlands
e-mail: P.P.Schot@uu.nl

P. de Voogt
KWR Watercycle Research Institute, Nieuwegein, The Netherlands

Institute for Biodiversity and Ecosystem Dynamics, University of Amsterdam, Amsterdam, The Netherlands
e-mail: W.P.deVoogt@uva.nl

A. van Wezel
Copernicus Institute of Sustainable Development, Faculty of Geosciences, Utrecht University, Utrecht, The Netherlands

KWR Watercycle Research Institute, Nieuwegein, The Netherlands
e-mail: Annemarie.van.Wezel@kwrwater.nl

© Springer International Publishing AG 2017
P. de Voogt (ed.), *Reviews of Environmental Contamination and Toxicology*
Volume 246, Reviews of Environmental Contamination and Toxicology 246,
DOI 10.1007/398_2017_10

1

Contents

Abbreviations

AU	Australia
BMDL	Benchmark dose
CA	Concentration addition
DE	Germany
DWD	Drinking Water Directive
EC50	Half maximal effective concentration
EU	European Union
GA	Glutaraldehyde
GWD	Groundwater Directive
IARC	International Agency for Research on Cancer
IEA	International Energy Agency
Koc	Soil organic carbon-water partition coefficient
LC-HRMS	Liquid chromatography high resolution mass spectrometry
Log Kow	n-octanol-water partition coefficient
MS	Mass Spectrometry
NOAEL	No adverse effect level
PBT	Persistent, bioaccumulative and toxic
PEC	Predicted environmental concentration
PNEC	Predicted no effect concentration
RfD	Reference dose
RQ	Risk quotient
TDS	Total dissolved solids
TDI	Tolerable daily intake
TTC	Threshold of toxicological concern
UK	United Kingdom
UO&G	Unconventional oil and gas
USA	United States of America
WFD	Water Framework Directive

1 Introduction

Unconventional oil and gas (UO&G) resources represent large volumes of hydrocarbons trapped inside relatively impermeable rock layers, making them more difficult to access than conventional resources (Elliott et al. 2017; Werner et al. 2015). These energy sources include oil and gas from shale formations, limestone, sandstone, and coal deposits. In order to reach these formations, wells need to be drilled several kilometers deep followed by additional horizontal drilling in order to cover a larger area (Jackson et al. 2013a). Large amounts of water (~90%) mixed with proppants (~9%) such as sand and chemical additives (~1%) (Vidic et al. 2013) are injected into the formations. For well injection in shale, some 8–19 million L of fracturing fluid is needed per well to perform one hydraulic fracture resulting in high loads of chemicals (King 2012). While hydraulic fracturing was already developed in the 1940s (Montgomery and Smith 2010), technological advancements in directional drilling and reservoir stimulation have made the extraction of unconventional energy sources economically viable since the past few decades (Jackson et al. 2013a). Hydraulic fracturing might help to secure energy needs and is sometimes considered as a step in the transition from coal to renewable energy production (Howarth et al. 2011). However, the extraction of unconventional hydrocarbons might also delay the development of renewable and sustainable energy policies and technologies, by competing for investments and deviating public and political attention (Howarth et al. 2011).

There is an increasing public and scientific concern about air, soil, and water contamination, due to the possible adverse health and environmental effects (Gordalla et al. 2013; Grant et al. 2015; Elliott et al. 2017). This paper focuses on the water system, where contamination might result both from chemical additives used in drilling and fracturing fluids and from components naturally present in the subsoil that are brought to the surface via drill cuttings, flowback, and produced waters (Vidic et al. 2013). These substances include heavy metals, radionuclides, brine, and hydrocarbons. The majority of studies on water contamination related to UO&G were carried out in the USA, i.e., Pennsylvania, Colorado, and Texas. Contamination after UO&G operations is reported in drinking water for manganese (Alawattegama et al. 2015), stray gas (Osborn et al. 2011), arsenic, selenium, strontium, and total dissolved solids (Fontenot et al. 2013; USEPA 2016). In groundwater samples after UO&G-related surface spills, benzene, toluene, ethylbenzene, and xylene exceeded drinking water guidelines (Gross et al. 2013), and brine contamination was reported (Preston and Chesley-Preston 2015). Water contamination can occur during gas extraction activities through surface and underground spills or leaks (Vidic et al. 2013). Wastewater disposal is a concern for water quality. In the USA, wastewater treatment plants are often used; however, they are not all well-equipped to efficiently treat the unconventional wastewaters.

This might, after disposal of the treated effluents, lead to surface water and shallow groundwater contamination (Ferrar et al. 2013; Butkovskyi et al. 2017).

Chemical risk assessment is used to allow safe use of chemicals in an array of sectors and is typically done per single compound and type of use, combining information on hazardous properties with expected exposures (Van Wezel et al. 2017). Environmental exposure scenarios are typically limited to surface water and relatively shallow groundwater. UO&G operations however have specific characteristics that might require adaptations of chemical risk assessment to properly assess the risks associated with these activities. The large number of chemicals involved might require further prioritization of these chemicals, e.g., with respect to risks for drinking water (Sjerps et al. 2016; Schriks et al. 2010a). Environmental fate processes including transformation might deviate from aboveground ones due to higher pressures and temperatures in the deep soil (Hoelzer et al. 2016), impacting chemical risk assessment.

Here, we identify and describe uncertainties and knowledge gaps of chemical risk assessment related to unconventional drillings, assess the available exposure models in relation to UO&G, and propose adaptations where necessary and possible including attention to monitoring practices. We discuss how chemical risk assessment in the context of UO&G differs from conventional chemical risk assessment and the implications for existing legislation concerning authorization of chemicals, unconventional drillings, and water quality.

2   Chemical Assessment

2.1   Chemicals Involved and Analytical Methods

The additives used in fracturing and drilling fluid include biocides, scale and corrosion inhibitors, oxygen scavengers, cleaners, gelling agents, friction reducers, iron controls, surfactants, cross-linkers, breakers, conditioners, and clay stabilizers. The number and volume of chemicals needed depend on the local subsurface conditions and chemical properties of the water used (Vidic et al. 2013). FracFocus Chemical Disclosure Registry (US Fracfocus 2016; Soeder et al. 2014) is a US database that contains over 1,100 different chemicals documented to be used during hydraulic fracturing so far, including a description of their purpose. Although registration is mandatory in several US states in order to get licenses, due to property rights and trade policies in many cases, exceptions are possible (Centner and O'Connell 2014; Maule et al. 2013). The quality of the entries in this database can be improved with regard to incorrectly reported CAS numbers, the use of different names for the same chemical, spelling mistakes, etc. In Europe, the International Association of Oil and Gas Producers introduced a registry still on a voluntary basis which for the moment only contains chemicals used in Polish unconventional drilling sites (NGS 2016). Poland as the most active European state in shale gas exploration and production provides a separate chemical registry

(OPPPW 2016). Local operators in other European countries also provide registries (NAM 2016; Cuadrilla 2016; ExxonMobile 2016).

Subsurface contaminants, i.e., heavy metals, radionuclides, salts, and hydrocarbons, are mobilized during drilling and hydraulic fracturing activities (Jackson et al. 2013b). Both drilling and hydraulic fracturing activities mobilize contaminants from the formation, and drilling mobilizes more diverse contaminants from overlying layers. There are currently no databases for subsurface contaminants, but many of the contaminants can be found in literature (Online Resource 1 – ESM 4; Abualfaraj et al. 2014; Alawattegama et al. 2015; Dahm et al. 2011; Ferrar et al. 2013; Fontenot et al. 2013; Grant et al. 2015; Gregory et al. 2011; Gross et al. 2013; Hayes 2009; Hayes and Severin 2012; Heilweil et al. 2015; Hildenbrand et al. 2015; Hladik et al. 2014; Lester et al. 2015; Maguire-Boyle and Barron 2014; Orem et al. 2014; Olsson et al. 2013; Osborn et al. 2011; Preston and Chesley-Preston 2015; Tang et al. 2014; Thacker et al. 2015; Warner et al. 2013a, b; Ziemkiewicz et al. 2014).

In view of the large number of chemicals that can be involved in hydraulic fracturing activities, there is a need for advanced analytical techniques to identify chemicals present in fracturing fluid, flowback and produced waters, shallow and deeper groundwaters, and treated wastewaters. For inorganic compounds, inductive coupled plasma mass spectrometry is generally used (Chapman et al. 2012; Strong et al. 2013; Ferrer and Thurman 2015b), although atomic absorption has also been utilized (Barbot et al. 2013; Ferrer and Thurman 2015b). Thermal ionization mass spectrometry may be used for radionuclide analysis (Chapman et al. 2012; Warner et al. 2013a; Ferrer and Thurman 2015b). Gas chromatography mass spectrometry (GCMS) may be used for organic volatile compounds, such as methane. For relatively polar organic compounds, high-resolution mass spectrometry (HRMS) may be used to perform both target and nontarget screening analysis (Hogenboom et al. 2009; Krauss et al. 2010; Ferrer and Thurman 2015a, b; Leendert et al. 2015; Schymanski et al. 2014a, b, 2015). Chemical MS analysis can provide quantification of single chemicals if standards are used and otherwise semi-quantification based on internal standard equivalents (e.g., Sjerps et al. 2016). MS analysis is only able to detect ionizable compounds, i.e., compounds that have at least one heteroatom (e.g., N, S, O, and P). Further details on sample preparation have been described by Ferrer and Thurman (2015b).

Nontarget screening differs from target screening in that it aims to detect all the substances present in a given sample, limited only by the analytical detection method (Müller et al. 2011). A list of 1,386 chemicals that might be expected in the UO&G water samples has been prepared using the aforementioned databases and literature (Online Resource 1 – ESM 1). The UO&G suspect list provides the compound names, the corresponding chemical formulas, CAS numbers, molecular weights, and their type of use (i.e., biocide, scale inhibitor, subsurface contaminant, etc.) where possible. Chemical formulas and molecular weights are based on Sjerps et al. (2016), PubChem, ChemSpider, and Toxnet. For the fracturing fluid additives, purposes and additive classes were indicated based on the US FracFocus database. The accurate masses in the UO&G suspect list can be used to compare with accurate masses of chemicals found in hydraulic fracturing-related water samples using liquid chromatography (LC)-HRMS suspect screening after which the identification can be further confirmed (Bletsou et al. 2015; Hug et al. 2014; Schymanski et al. 2014b). For this purpose the list needs to be reduced to only include ionizable compounds that can be detected with LC-HRMS (Online Resource 1 – ESM 2; Sjerps et al. 2016). The list is made up of 21% subsurface contaminants and 79% fracturing fluid additives (Fig. 1). Most of the additives have multiple functions (27%), followed by tracers (21%) and corrosion inhibitors (8%). The other purposes (base fluid, biocide, breaker, clay control, cross-linker, friction reducer, gelling agent, iron control, proppant, scale inhibitor, surfactant) all fall below 5%. The detailed composition of the 5% other can be found in Online Resource 1 – ESM 5a. Some compounds were registered as "additional ingredients," without further detail on their specific function. Only 44% and 52% of the compounds of the complete and the reduced lists, respectively, are regulated in the EU (Table 1; ECHA 2017).

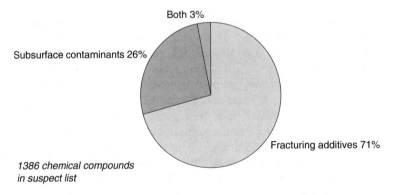

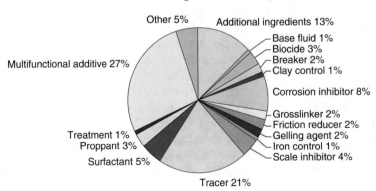

Fig. 1 Composition of UO&G suspect list, including an overview of the fracturing fluid additive purposes (Online Resource 1 – ESM 5b and 6a)

Table 1 Suspect list compounds regulated in Europe (Online Resource 1 – ESMs 1 and 2)

	Complete list	EU regulated	LC-HRMS analyzable List	EU regulated
Total suspects	1,386	606 (44%)	462	242 (52%)
Fracturing additives	1,043	473 (45%)	304	184 (61%)
Subsurface contaminants	325	133 (41%)	158	58 (37%)

This means that if unconventional hydrocarbons were to be extracted on a large scale in Europe, there would either be a limited number of compounds to choose from for the composition of the fracturing fluid or more work needs to be done in regulating these compounds.

Chemical mixture effects cannot be covered by chemical analysis alone (Escher and Leusch 2012). Effect-directed analysis, using in vitro bioassays, complements the chemical MS analysis and provides insight into all bioactive chemicals (Kolkman et al. 2013). In vitro analysis provides a good base in toxicological endpoints for health risks (Arini et al. 2016; Escher et al. 2013; Leusch et al. 2017; Murk et al. 2002; Nelson et al. 2007; Poulsen et al. 2011; Brand et al. 2013; Schriks et al. 2010b). More specifically, UO&G-related chemicals have been associated with adverse developmental and reproductive effects (Webb et al. 2014) and endocrine-disrupting effects (Kassotis et al. 2013, 2015, 2016a, b), so a hormonal-based and/or a reproductive-/developmental-based assessment of water mixtures related to UO&G operations could be of importance for detecting contamination due to UO&G activities.

2.2 Concentrations and Loads in UO&G-Related Waters

The reported concentrations of chemicals used and mobilized during UO&G operations in related water for shale gas were recently reviewed by Annevelink et al. (2016). Here we actualize and extend this information for other types of unconventional hydraulic fracturing (Table 2). Figure 2a, b presents a selection of compounds measured in all three matrices, i.e., surface water or groundwater, flowback or produced water, and wastewater. An overview of reported concentrations for all compounds can be found in Online Resource 2 (ESM1). Of all of the compounds analyzed in literature, most were analyzed in flowback and produced waters (72%), followed by fracturing fluid (59%) and surface water and groundwater (59%). In addition, most compounds were measured in relation to shale gas operations (84%), followed by coal-bed methane (53%), and only very few were measured for tight gas and conventional gas (11%; Online Resource 2 – ESM 3).

Baseline data is often not ensured, which complicates conclusions on the significance of UO&G as source of contamination (Lange et al. 2013). Generally, the highest concentrations can be found in flowback and produced waters, whereas

Table 2 Overview of matrices, formations, locations, and compound types analyzed in literature

Water type	UO&G type	Location	Compounds measured	Reference
Fracturing fluid	Shale gas	Pennsylvania (US)	Inorganics and organics	Ziemkiewicz et al. (2014)
	Shale gas	Pennsylvania and West Virginia – Marcellus (US)	Inorganics and organics	Hayes (2009)
	Shale gas	Barnett and Appalachian shale (US)	Inorganics and organics	Hayes and Severin (2012)
Flowback/ produced	Shale gas	Pennsylvania, New York, West Virginia (US)	Inorganics and organics	Abualfaraj et al. (2014)
	Shale gas	Pennsylvania and West Virginia – Marcellus (US)	Inorganics and organics	Hayes (2009)
	Shale gas	Barnett and Appalachian shale (US)	Inorganics and organics	Hayes and Severin (2012)
	Shale gas	Colorado (US)	Inorganics and organics	Lester et al. (2015)
	Shale gas	Pennsylvania, Texas, New Mexico (US)	Inorganics and organics	Maguire-Boyle and Barron (2014)
	Shale gas	Cappeln, Damme, Buchhorst (DE)	Inorganics	Olsson et al. (2013)
	Shale gas	Pennsylvania, Indiana, Kentucky (US)	Organics	Orem et al. (2014)
	Shale gas	Pennsylvania (US)	Inorganics	Warner et al. (2013a)
	Shale gas	US	Inorganics and organics	Thacker et al. (2015)
	Shale gas	Marcellus gas wells (US)	Inorganics and organics	Ziemkiewicz et al. (2014)
	Shale gas	Marcellus shale (Western Pennsylvania – US)	Inorganics	Gregory et al. (2011)
	Coal-bed methane	Illinois, Alabama, Wyoming, Montana, North Dakota (US)	Organics	Orem et al. (2014)
	Coal-bed methane	Colorado, Wyoming, New Mexico (US)	Inorganics and organics	Dahm et al. (2011)
	Coal-bed methane	US	Inorganics	Thacker et al. (2015); Alley et al. (2011)
	Tight gas	US	Inorganics	Thacker et al. (2015); Alley et al. (2011)
	Conventional gas	US	Inorganics	Thacker et al. (2015); Alley et al. (2011)

(continued)

Table 2 (continued)

Water type	UO&G type	Location	Compounds measured	Reference
Surface water and groundwater	Shale gas	Pennsylvania (US)	Inorganics	Alawattegama et al. (2015)
	Shale gas	Texas (US)	Inorganics and organics	Fontenot et al. (2013)
	Shale gas	Pennsylvania (US)	Inorganics and organics	Ferrar et al. (2013)
	Shale gas	Pennsylvania, New York (US)	Inorganics and organics	Osborn et al. (2011)
	Shale gas	Montana, North Dakota (US)	Inorganics	Preston and Chesley-Preston (2015)
	Shale gas	Pennsylvania (US)	Inorganics	Warner et al. (2013a)
	Shale gas	Arkansas (US)	Inorganics	Warner et al. (2013b)
	Shale gas	Marcellus gas wells (US)	Inorganics and organics	Ziemkiewicz et al. (2014)
	Shale gas	Pennsylvania (US)	Organics	Heilweil et al. (2015)
	Shale gas	Pennsylvania (US)	Inorganics	Grant et al. (2015)
	Shale gas	Marcellus shale (Colorado, US)	Organics	Gross et al. (2013)
	Shale gas	Barnett shale (Texas, US)	Inorganics and organics	Hildenbrand et al. (2015)
	(Un)conventional gas	Pennsylvania, Colorado, Maryland, Virginia (US)	Organics	Hladik et al. (2014)
	Coal-bed methane	Queensland (AU)	Inorganics	Tang et al. (2014)

US United States, *DE* Germany, *AU* Australia

the lowest concentrations are found in surface/shallow aquifers (Fig. 2a, b, Online Resource 2 – ESM 2a). Most studies focus on chemicals present in wastewater, followed by surface and shallow aquifers. Only few studies (Ziemkiewicz et al. 2014; Hayes 2009; Hayes and Severin 2012) measured compounds in fracturing fluid. Although contamination in one matrix can influence a subsequent matrix, no study analyzed the whole cycle from fracturing fluid, flowback and produced water, and surface water and groundwater related to UO&G operations.

The majority of available studies focus on the US UO&G operations: only two studies are available from Australia and Germany (Table 2). Most studies reported concentrations of chemicals in water from shale gas activities (Table 2); only few studies (Dahm et al. 2011; Orem et al. 2014; Thacker et al. 2015; Alley et al. 2011) relate to coal-bed methane, tight sand gas, or conventional gas-related activities.

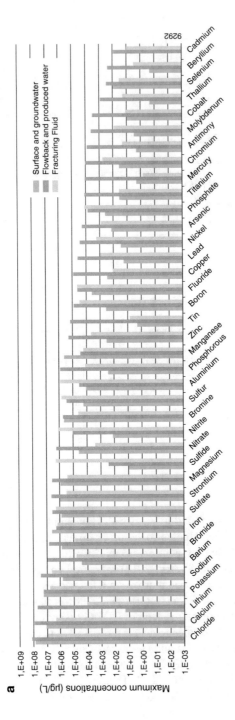

Fig. 2 (**a**) Maximum reported concentrations (µg/L) for a selection of inorganic compounds measured in surface water and groundwater, flowback and produced waters, and fracturing fluid, based on sources in Table 2 (Online Resource 2 – ESM 2a); no bar represents a compound measured below the detection limit. (**b**) Maximum reported concentrations (µg/L) for a selection of organic compounds measured in surface water and groundwater, flowback and produced waters, and fracturing fluid, based on sources in Table 2 (Online Resource 2 – ESM 2a); no bar represents a compound measured below the detection limit

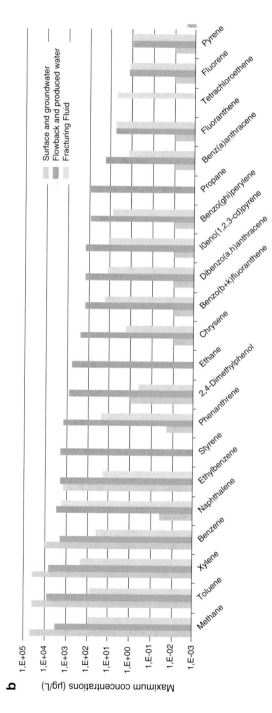

Fig. 2 (continued)

This attention toward shale gas is explained by its high production; in IEA member countries excluding China and Albania due to missing data, 58% of the UO&G relates to shale, while 34% and 8% relate to tight sand gas and coal-bed methane, respectively (IEA 2017).

Generally in the found literature, target screening was used, probably missing contaminants of concern. The majority of the chemicals from the UO&G suspect list are not measured in the available literature, for only 4% information is available (Online Resource 1 – ESM 6b). So, because of this scarce data availability, it is still difficult to address the risks of UO&G for water systems (Jackson et al. 2013b).

2.3   Monitoring and Recommendations

Routine surface water and groundwater monitoring practices are currently not designed to consider specific risks associated with UO&G operations. The Water Framework Directive (WFD; 2000/60/EC), including the Groundwater Directive (GWD; 2006/118/EC) and the Drinking Water Directive (DWD; 98/83/EC), outlines requirements for monitoring of surface water and groundwater in Europe, related to the physical, chemical, and biological water characteristics. The intensity of monitoring effort, however, varies hugely among water bodies (Malaj et al. 2014). Generally, deep (>100 m) groundwater is not monitored, while anthropogenic effects on groundwater can remain for decades (Sophocleous 2002). General water quality monitoring is not specified to chemicals related to UO&G practices or the transformation products that may be formed during the hydraulic fracturing process (Hoelzer et al. 2016). Furthermore, Harris et al. (2016) highlight the difficulty in detecting spill events: the distance travelled by a contaminant strongly affects its concentration in surface water, potentially leading to very subtle concentration changes depending on the monitoring location. Burton et al. (2016) found that beryllium is a good indicator for evaluating UO&G-related impacts on regional groundwater quality, in the Barnett Shale formation. Such indicators could also be determined for other important UO&G formations.

We recommend regular long-term monitoring of the whole UO&G-related water cycle including baseline data in the surroundings of UO&G operations, focusing on the specific persistent and mobile chemicals used, their transformation products, and the chemicals originating from the subsoil (cf. Vidic et al. 2013). Baseline data is needed in order to monitor changes resulting from the UO&G activities. Regular long-term monitoring ensures the timely detection of leaks after the well has been abandoned so that remediation efforts can quickly be put in place (Patterson et al. 2017; Maloney et al. 2017). Such monitoring programs could include effect-directed analysis (Brack et al. 2017; Venkatesan and Halden 2015). A significant increase in monitoring data related to a variety of geological and hydrological situations will improve our understanding of groundwater contamination related to UO&G operations (Soeder 2015). Indeed, the Commission Recommendation

(2014/70/EU) on hydraulic fracturing does provide UO&G-tailored provisions about, inter alia, risk assessments, monitoring, and baseline studies, but due to its non-binding nature, its effectiveness yet remains uncertain. Therefore, in order to better implement the above recommendations and to increase compliance potential and uniformity in practices, it should be taken into consideration to include them into the general framework of EU water legislation, preferably building upon and further specifying existing soft law documents, such as Recommendation 2014/70/EU.

3   Assessment of Hazardous Properties of the Chemicals

After chemical analysis, not only a list of chemicals present in UO&G-related waters is obtained but also their concentrations. These include chemicals with high and low toxicological concern for humans and the environment. In the EU, industrial chemicals and biocides used in the fracturing fluids are registered and authorized via the REACH regulation (EC 1907/2006) and the Biocidal Products Regulation (EU 528/2012). REACH defines chemicals of concern as persistent, bioaccumulative, and toxic (PBT; EC 1907/2006 – Annex XIII REACH). For water contamination, relatively hydrophilic compounds are more problematic compared to bioaccumulative chemicals (Reemtsma et al. 2016, Westerhoff et al. 2005, Sjerps et al. 2016). Chemicals that have carcinogenic, mutagenic, and toxic reproductive properties are defined in REACH as chemicals of concern (CMR; EC 1907/2006 – Annex VI of CLP regulation). Water quality requirements are presented in the European WFD, including the GWD, and in the DWD. However, as valid for chemicals in general, most of them used in drilling and fracturing fluid are not regulated by drinking water or water quality regulations other than in a generic sense. For non-regulated chemicals provisional water quality limits based on substance specific toxicity data can be used to estimate safe exposure levels (Schriks et al. 2010a). In absence of the latter, the more conservative and precautionary concept of thresholds of toxicological concern (TTC) can be used (Mons et al. 2013; Kroes et al. 2005) as based on toxicological data from a representative set of compounds. For drinking water, this conservative value is set at 0.1 and 0.01 µg/L for organic compounds and for genotoxicant and endocrine active chemicals, respectively (Mons et al. 2013). Provisional water quality limits, based on chemical specific toxicity data, can be orders of magnitudes less stringent (Schriks et al. 2010a).

The toxicological information available regarding chemicals related to UO&G operations is generally limited. Elliott et al. (2017) noted that more than 80% of the compounds of a list of UO&G-related water contaminants and air pollutants were not evaluated as to their carcinogenicity by IARC. In the USA, 87% of all chemicals used in fracturing fluids are not assessed for chronic toxicological effects (Yost et al. 2016). Furthermore, Shonkoff et al. (2014) identified the need for more epidemiological studies to assess UO&G water pollution in combination with adverse health effects among people living close to UO&G operations. This is corroborated by Stringfellow et al. (2014) who collected physicochemical and toxicological data for 81 fracturing fluid additives and did not find toxicological information for 30 of these chemicals.

Human and ecotoxicity are evaluated based on different endpoints, including acute toxicity tests, but also chronic endpoints such as carcinogenicity, mutagenicity or development, and reproduction. The available toxicological information varies strongly among chemicals, e.g., regarding the number of species tested, the endpoints studied, the duration, and pathways of exposure. Not all toxicological data available in the dossiers for registration and authorization is available in open literature and databases. If hardly any toxicological information is available, risks can be estimated using QSAR or read-across approaches (Gramatica 2007; Lee and Von Gunten 2012; Kühne et al. 2013; Scholz et al. 2013). There are several databases available for toxicological information on chemicals, e.g., Toxnet, the Hazardous Substances Data Bank (HSDB), the International Toxicity Estimates for Risk (ITER), the International Programme on Chemical Safety INCHEM Database, and OECD's eChemPortal. These databases include information from US National Toxicology Program (NTP). The International Agency for Research on Cancer (IARC) classifies carcinogenic compounds. Remaining information can be found in peer-reviewed scientific literature, using, e.g., scopus or google scholar search engines, by using keywords "toxicity, reference dose (RfD), acceptable daily intake (ADI), tolerable daily intake (TDI), no adverse effect level (NOAEL), and benchmark dose lower confidence limit for a 10% response (BMDL10)" combined with the chemical name or CAS number. The OECD QSAR toolbox can be used to fill in the gaps in (eco)toxicity data.

Table 3 provides an overview of (eco)toxicological data for a selection of five chemicals with high concentrations in surface water and groundwater, fracturing and produced water, and fracturing fluid. Carcinogenic data, genotoxic data, developmental/reproductive toxicity, NOAEL, TDI, and reference doses were available for all chemicals. Benzene has the lowest reference dose, followed by naphthalene, toluene, ethylbenzene, and lastly xylene. The lowest EC50 value was taken for every chemical in order to evaluate the worst-case scenario. BMDL data was only available for a few chemicals. These results show that for these very well-known chemicals, toxicological data is generally available. EPA's Ecotox database is a good source for toxicological data in terms of completeness. However, for some toxicological data, the information is scattered over different sources.

Although concentrations of chemicals in UO&G-related wastewater generally exceed TTC values (Fig. 2), this only becomes problematic if these untreated waters come into contact with freshwater reservoirs. Almost half of the organic compounds analyzed in surface water and groundwater exceed TTC values, indicating that further detailed risk assessment is needed and risks related to the extraction of UO&G cannot be waived beforehand. Assessing the toxicological effect of mixtures is important for unconventional drilling activities because of the large variety of chemical additives or subsurface contaminants involved (Riedl et al. 2013). It is generally accepted that the concept of concentration addition (CA) can be used as a precautious first tier to assess mixture toxicity (Backhaus and Faust 2012; SCHER

Table 3 (Eco)toxicological data for five organic compounds with high concentrations measured in surface water and groundwater, flowback/produced water, and fracturing fluid and their respective risk quotients, including the mixture toxicity assessment (Total RQ)

Compound	CAS	Toxicity Carcinogenic	Genotoxic	Developmental/ reproductive	NOAEL (mg/kg bw/day)	RfD (mg/kg bw/day)	TDI (mg/kg bw/day)	BMDL (mg/kg bw/day)	Ecotoxicity EC50 algae (µg/L)	EC50 Daphnia (µg/L)	EC50 fish (µg/L)	RQ = PEC/ PNEC
Toluene	108-88-3	3: Not classifiable as to its carcinogenicity to Humans[a]	Yes (Animal)[a]	Yes (Animal)[b]	312[c]	0.08[d]	0.08[d]	238[d]	9,400[e]	6,000[e]	1,650[e]	43.03
Benzene	71-43-2	1: Carcinogenic to Humans[a]	Evidence (Human)[f]	Evidence (Animal)[g]	50[h]	0.004[d]	0.004[d]	1.2[d]	29,000[e]	9,230[e]	1,740[e]	7.47
Ethyl-benzene	100-41-4	2B: Possibly carcinogenic to humans[a]	No evidence[i]	Evidence (Animal)[j]	136[k]	0.1[d]	0.097[k]	48[l]	1,340[e]	1,810[e]	4,200[m]	6.72
Xylene	1330-20-7	3: Not classifiable as to its carcinogenicity to Humans[a]	No evidence[d]	Yes (Animal)[d]	179[d]	0.2[d]	0.15[k]		3,000[e]	76,201[e]	4,000[e]	13.00
Naphthalene	91-20-3	2B: Possibly carcinogenic to humans[a]	Not determined[f]	Yes (Animal)[n]	71[d]	0.02[d]	0.04[k]		2,820[e]	690[e]	1,600[o]	0.06
Total RQ												70.28

EC50 half maximal effective concentration, *NOAEL* no observed adverse effect concentration, *NOAEL* no observed adverse effect level, *RfD* reference dose, *TDI* tolerable daily intake, *BMDL* benchmark dose, *RQ* risk quotient, *PEC* predicted environmental concentrations. The maximum concentrations reported in surface water and groundwater are used here, *PNEC* predicted no effect concentration

Sources: [a]IARC (2017), [b]ACGIH (2013), [c]NTP (1990), [d]US EPA IRIS (2017), [e]US EPA Ecotox (2017), [f]EFSA (2017), [g]USEPA (2006), [h]Van Herwijnen and Vos (2009), [i]Zhang et al. (2010), [j]DHHS ATSDR (2010), [k]Baars et al. (2001), [l]Mellert et al. (2007), [m]Smit and Verbruggen (2012), [n]Navarro et al. (1992), [o]DeGraeve et al. (1982)

SCENIHR SCCS 2012). A mixture made up of the organic compounds among the five highest reported concentrations in surface water/groundwater, flowback/production water, and fracturing fluid represents a high risk quotient (RQ) of 70.28 (Table 3). Except for naphthalene, all of the chemicals have RQs higher than 1, indicating a risk, toluene showing the highest RQ. This calculation is based on a realistic worst-case scenario, with maximum reported concentrations in surface water and groundwater, as well as minimum reported EC50 used. An assessment factor of 1,000 was applied. A comparable PEC-/PNEC-based risk assessment has also been conducted by Butkovskyi et al. (2017) for a selection of commonly used organic chemicals in UO&G-related operations. They identified a number of potentially harmful compounds from the shale formation (polycyclic aromatic hydrocarbons, phthalates), from the fracturing fluids (quaternary ammonium biocides, 2-butoxyethanol), and as a result of downhole transformation (carbon disulfide, halogen).

4 Exposure

Chemicals having hazardous properties will only lead to risks if humans or ecosystems are exposed. To estimate exposure, both the emission rates and the chemical fate processes are important. However, no specific exposure scenarios toward groundwater aquifers exist for UO&G-related activities. Human errors in various stages of the life cycle of UO&G production play an important role in these exposure scenarios. Surface water and groundwater contamination might occur at the surface via accidental spills or in the subsoil via leaks (Gordalla et al. 2013) due to structure integrity problems or human errors. Surface spills can affect surface waters and shallow aquifers via infiltration or direct leaching. Underground leaks can affect aquifers via migration through artificial and/or natural faults and fractures. Spills and/or leaks related to human error occur mainly due to insufficient cementing, leaking connectivity, and blowouts (Table 4). Accidental surface spills do not seem to impact groundwater systems but do however impact surface waters (Harkness et al. 2017). Emission of the UO&G-related chemicals highly depends on failure probabilities. These can be estimated for surface or near-surface spills and leaks by considering publicly available data on spill occurrences and released volumes from the USA. Failures and consequently leak volumes occurring deep underground could be underreported in these databases due to the lack of monitoring at these depths, even though monitoring of pressure decreases might be ensured. Due to the scarcity of UO&G-related operations in the EU, there is little information available on spill frequencies and volumes for the European situation, meaning that related studies must rely on US databases. Identification of failures related to human error and their frequencies and spill/leak volumes from 2010 to 2015 are summarized in Table 4. The data is based on publicly available US governmental databases (NRC 2014; COGC 2014; OCD 2014; PADEP 2014; RRC 2014). The failure probability ranges were determined by relating the number of incidents to

Table 4 Spill/leak probabilities and spill volume estimates based on US publicly available databases (2010–2015): National Response Center (NRC), Colorado Oil and Gas Commission (COGC), Oil Conservation Division (OCD), Pennsylvania Department of Environmental Protections (PADEP), and Railroad Commission Texas (RRC)

Contamination pathway	Fluid released	Frequency (%/well/year)	Average spill volume (m^3)
Surface spill	Drilling mud	0.005–2.8	294 ± 185.7
	Fracturing fluid	0.02–0.1	24 ± 28
	Produced water	0.02–4.4	12 ± 29.1
	Oil-based fluid	0.05–2.8	1 ± 6
Blowout	Drilling mud	0.004	185 ± 256
	Produced water	0.0002–0.01	3,206 ± 7,843
	Oil-based fluid	0.002–0.01	49 ± 243
Leaking connectivity	Drilling mud	0.01	43 ± 50
	Produced water	0.2	12 ± 26
	Oil-based fluid	0.1	6 ± 14
Corroding well casing	Oil-based fluid	0.05–0.7	9 ± 20
	Drilling mud	0.001–0.004	4 ± 4
	Produced water	0.002–1	11 ± 41
Insufficient cementing	Not specified	1.6	Not specified

the active wells in the relevant time frame and area for each database. The spill/leak volumes were calculated by multiplying the average reported spill volume with the minimum and maximum failure probability. Insufficient cementing (irrespective of the fluid) appears to be the main reason for the occurrence of failures, followed by oil-based fluids released due to leaking connectivity. Blowouts are the least frequent reason for failures; however, the associated average spill volume is the highest of all the studied mechanisms. Leaking connectivity and corroding well casing are the contamination pathways that result in the least fluid spilled. Furthermore, O'Maloney et al. (2017) looked into spill frequencies and volumes relating to the type of fluid for different US member states. They found that the main fluids of concern are UO&G-related wastewater, crude oil, fracturing fluid, and drill waste. While crude oil is responsible for the highest frequency of spills, it has the second lowest related spill volume. Fracturing fluid has the lowest spill frequency but the highest spill volume. Patterson et al. (2017) mainly studied the US spill rates related to the well life and their causes. 2–16% of wells report yearly spills, and 75–94% of spills occur in the first 3 years of well life. Storage and pipeline transport were responsible for 50% of the reported spills.

During wastewater production, most of the injected fracturing fluid (92–96%) remains in the subsurface formation (Kondash et al. 2017). Most of this water resides in the shale matrix, and only a small portion goes into surrounding fractures (O'Malley et al. 2015). The chemical fate of the UO&G-related chemicals in the water system depends on their physicochemical characteristics and the circumstances in the matrix. For inorganics pH, oxidation state, presence of iron oxides, soil organic matter, cation exchange capacity, and major ion chemistry are important. Their transport is governed by physical flow processes (advection and dispersion), sorption, and precipitation (USEPA 2015). Effects and impacts of organic substances may be altered by chemical degradation and other loss processes, e.g., sorption, oxidation, volatilization, etc. (USEPA 2015; Escher and Schwarzenbach 2002; Schnoor 1996). Some parameters are especially important when looking into chemical fate and transport in water systems. The n-octanol-water partition coefficient (log K_{ow}) relates to the accumulation in biological organisms and sorption to organic matrices as expressed by the soil organic carbon-water partition coefficient (K_{oc}). Degradation half-lives describe the persistence of the chemicals and their susceptibility to (bio) degradation. Furthermore Henry's law constant describes volatilization. Again, data-bases are available that list or estimate these parameters such as EPISUITE, SPARC, or OECD's eChemPortal. For predicted parameters inorganic compounds fall outside the estimation domain (Gouin et al. 2004), and estimations might be inaccurate.

Current conceptual box models such as QWASI (Mackay et al. 2014) or SIMPLEBOX (Hollander et al. 2016) describe chemical fate in surface waters or shallow groundwater. Surface exposures can be modeled, but they do not take into account specific routes involved in UO&G operations. As described above, UO&G-related exposures can originate from multiple sources, making it important to consider all the different exposure paths. Furthermore, a chemical deep under-ground will be subjected to high temperature (up to 200°C), high pressure (above 10 MPa), and high salinity (TDS: 100,000–300,000 mg/L), which might alter chemical behavior (Kahrilas et al. 2016). Increasing temperatures generally lead to higher solubility and reduced sorption (Huang 1980; Mackay 1980). Higher temperatures will also affect the degradation rate and up to a maximum temperature reduce half-lives (Klein 1989). There is not much information on chemical behavior under these conditions, except for glutaraldehyde (GA; Kahrilas et al. 2014, 2016). Kahrilas et al. (2016) found for this one chemical that the main conditions impacting chemical fate are temperature, pH, and salinity, whereas pressure was not found to be important. High temperatures and/or alkaline pH caused GA to rapidly autopolymerize and eventually precipitate, whereas high salinity inhibited GA transformation.

5   Discussion and Conclusion

5.1   Reducing Risk by Adequate Wastewater Management and Technological Innovation

Wastewater management is a critical stage of the water life cycle of UO&G production that can lead to environmental contamination (Camarillo et al. 2016; Kondash et al. 2017). Therefore, proper management is crucial in order to reduce the risks associated with UO&G operations. Wastewater in the USA is typically made up of 4–8% of injected fracturing fluid and 92–96% of naturally occurring formation brines, whereby 20–50% of the total produced wastewater is generated during the first 6 months of an active well (Kondash et al. 2017). Waste from drilling and fracking operations is typically stored in open air impoundments or in closed containers, to either be recycled for reuse or to be disposed via thermal treatment (incineration, pyrolysis, gasification), bioremediation, composting, or deep-well injection (Zoveidavianpoor et al. 2012; Boschee 2014; Camarillo et al. 2016; Butkovskyi et al. 2017). Wastewater management plans depend on regulations, cost, technology performance, location, and disposal alternatives (Gregory et al. 2011). Reinjection options are the cheapest and therefore also the most used management possibility, depending on a state's regulations (Clark and Veil 2009). However, due to the small number of adequate disposal wells in the USA and increasing stress on water availability in some areas, the need for wastewater reuse becomes more urgent. Consequently, the need for adequate treatment facilities gains importance, although the treatment technologies are challenged by the high salt content. Treated wastewater could be applied to roads for deicing and dust-suppression purposes (Warner et al. 2013a). Camarillo et al. (2016) and Butkovskyi et al. (2017) thoroughly describe the problems encountered related to wastewater treatment and possible solutions. More technological advancements are needed in wastewater treatment in order to promote reuse through efficient contaminant removal and cost reduction (Butkovskyi et al. 2017). Moreover, geotextiles and geosynthetics can be used at drilling sites to control surface failure probabilities and contamination effects, after spills have occurred (ter Heege et al. 2014). These are permeable fabrics, which are able to separate, filter, reinforce, protect, and drain wastewater from the surrounding environment. These products are readily available, but their use is not widespread (ter Heege et al. 2014). Underground leaks can be controlled by minimizing upward migration of contaminants, by keeping a minimum distance of 1.6 km between the bottom of the aquifer and the horizontal drill line (US EPA 2015).

5.2 Identified Knowledge Gaps and Uncertainties

There are several knowledge gaps and uncertainties that have been identified relating to the lack of evidence-based research, monitoring guidelines, available information on compounds, environmental fate models, and legislation. The number of publications that chemically characterize UO&G-related water systems is limited. Most of the studies focus on shale gas in the USA and only a select few on coal-bed methane and tight sand gas. There is also limited research on chemical risk assessment of conventional oil and gas recovery (Alley et al. 2011; Thacker et al. 2015; Afenyo et al. 2017). This is important for comparison reasons and would provide insight into whether the risks related to unconventional activities are in fact higher than those related to conventional activities. Chemical characterization studies of UO&G-related waters are generally based on target screening and do not consider the whole UO&G-related water cycle (i.e., fracturing fluid, flowback/produced water, groundwater). Additionally, there are no databases on subsurface contaminants that may be mobilized during UO&G operations, and the chemical registration of additives can be improved in FracFocus (USA). However, literature on subsurface contamination can be found. Registration of UO&G fracturing fluid chemicals in the EU is still on a voluntary basis. Furthermore, the guidelines for water quality monitoring are not specified for adequate risk assessment for unconventional operations. As long as UO&G-specific guidelines have not been set, an effect-directed assessment may provide a better insight into the risks to environmental or human health of UO&G-related waters. Moreover, despite "good intentions" such as the Commission Recommendation 2014/70/EU, baseline monitoring is generally not ensured, and due to limited access to the deep underground and high costs, there is limited data available on deep underground failures, which makes it difficult to assess failure probabilities and ultimately the associated risk. In addition, only a small proportion of chemicals used in fracturing fluid has been evaluated as to their chronic toxicity, meaning that regulation and authorization of chemicals need to be updated. Moreover, there is limited information available on changes in chemical behavior under downhole conditions, which can be important when assessing the environmental fate of chemicals accidentally released into the deep underground. As a result the available environmental fate models do not allow to determine environmental fate in such a situation.

Also research in the field of risk assessments from a regulatory perspective shows important gaps. Environmental legislation is found at various institutional levels, including the international, regional (e.g., EU), national/federal, and subnational (e.g., state, supra-local, and local) levels. Neither at the international level nor at the US federal and the EU levels, specific regulations for UO&G-related activities are in place to protect environmental and human health (Geraets and Reins 2016; Centner and Petetin 2015; Lange et al. 2013). Instead, both in the EU and the USA, there is a vast body of general legislation that is or can be applicable to UO&G operations. However, at the US federal level, the oil and gas industry, mainly on the basis of the Energy Policy Act of 2005, benefits from exemptions from several major federal environmental statutes, including the Clean Air Act, the Clean Water Act, and the Safe Drinking Water Act (Brady and Crannell 2012). As a result and despite attempts of the US Congress to pass the Fracturing Responsibility and Awareness of Chemicals Act in 2013, operators seeking to conduct hydraulic fracturing are excluded from permitting well construction and chemical disclosure rules at the federal level (Grant 2016). Thus, most relevant regulation of UO&G activities in the USA is to be found at the state level (Brady and Crannell 2012). Indeed, some states (e.g., Pennsylvania and Texas) have adopted general and/or specific regulations on UO&G activities, including risk assessment, permitting systems, rules aiming at groundwater protection, requirements for drilling, well construction (e.g., casing, cementing) and well control, monitoring, and reporting arrangements, as well as obligations to disclose information about chemicals used during the fracturing process (Polishchuk 2017; Grant 2016; Roberson 2012).

These regulations can be considered explicit, detailed, and well developed, albeit in some cases unclear and poorly disclosed (Polishchuk 2017).

At the EU level, the current legal framework governing UO&G activities mainly consists of general environmental principles, directives, and regulations (Vos 2014). Key environmental principles, listed in Article 191(2) of the Treaty on the Functioning of the European Union, are to be at the basis of the environmental legislation and policies of the EU and its member states; these are the precautionary principle (Reins 2014a), the prevention principle (Fleming and Reins 2016), the principle that environmental harm should be rectified at source, and the polluter pays principle. Relevant directives and regulations are, inter alia, the Environmental Impact Assessment Directive (2011/92/EU), the WFD, the GWD, the DWD, the Mining Waste Directive (2006/21/EC), the Hydrocarbons Directive (94/22/EC), and the REACH Regulation and the Biocidal Products Regulation (Polishchuk 2017; Kevelam 2015; Vos 2014). However, REACH and the Biocidal Products Regulation provide specific risk assessments, but not all relevant substances are assessed, whereas other directives have included risk assessments as part of a licensing procedure or the environmental impact assessment, but these have not specifically been designed for assessing the risks related to the different stages of UO&G operations. Indeed, the European Commission has issued UO&G-tailored recommendations (Commission Recommendation 2014/70/EU), providing minimum principles or guidelines regarding, inter alia, strategic planning and assessing environmental impacts, baseline studies, installation design and construction, operational aspects, and monitoring. However, apart from its non-binding nature, it remains uncertain whether these recommendations will be effective in increasing uniformity and environmental/human health protection. At the point of, for instance, risk assessment it only refers to the general EU legislation mentioned above. At other points, this recommendation provides more tailored provisions, but these are still very generally formulated, leaving much room for interpretation. Nonetheless, formulated as an invitation to member states, the Recommendation should also be seen as an important step in the development of a more effective regulation of UO&G activities in the EU.

Lastly, although there is a large body of literature about EU environmental principles, directives, and regulations in general, there is hardly any literature to be found applying these to UO&G activities (Vos 2014). Also at the level of the member states, specific UO&G regulations are mostly absent (e.g., the Netherlands; Kevelam 2015; Brans and Van den Brink 2014) or in a very early and rudimentary stage of development (e.g., Poland; Polishchuk 2017; Atkins 2013), which means that UO&G activities are mostly regulated through general environmental, planning, and mining legislation. Moreover, legal literature on dealing with known and uncertain risks does not focus specifically on UO&G activities but mostly on uncertain risks from an economic and governance perspective (Randall 2011) or from a general liability perspective or on specific aspects such as contract law (De Jong 2013, 2016; Pereira et al. 2016).

5.3   Implications for Risk Assessment and Legislation

EU and member states' "UO&G regimes" can be considered highly fragmented and at some points unsuitable and/or incomplete, particularly when it comes to regulating and monitoring the environmental risks of substances used in fracturing fluids or transformation products (Reins 2014b; also see Sects. 2.3 and 3). Therefore, further regulations should be implemented after the development of a better understanding of environmental risks related to UO&G operations (Gordalla et al. 2013; Reins 2017). For these regulations, the introduction of the provision of mandatory location-specific and phase-specific risk assessments by potential hydraulic fracturing operators is recommended as a prerequisite for a permission to conduct UO&G activities. In the UK, such a system already exists (Prpich et al. 2016). In this respect, a physics-based approach to detect and evaluate the possible risks for groundwater contamination, such as characterization of the system and migration pathways, could be implemented and used to compare risks between different sites to choose from (Lange et al. 2013). A spatial analysis method, mapping and quantifying the environment and population at risk, could provide further insight for risk assessment (Meng 2015). Changing industrial practices have the potential to modify behavior and fate of the compounds involved and should be considered (Goldstein et al. 2014). An evaluation of the risks for every stage of the hydraulic fracturing life cycle would allow for the prioritization and development of adequate management plans (Torres et al. 2016). Also social acceptance and perceptions from involved groups should be considered (Torres et al. 2016). The political decisions on unconventional drillings are not only based on scientific research but also influenced by the public, which means that the perception of risk by the general public is important for regulation. Familiarity with the process and trust are the two main factors influencing social beliefs (Wachinger et al. 2013). However, a recent study found that half of the people questioned were not familiar with the process of hydraulic fracturing, and the natural gas industry was considered an untrustworthy source of information (Theodori et al. 2014). A better long-term decision making procedure could be based on an iterative process where concerned parties come together to deliberate on risks, resulting in an improved transparency and understanding by the public on UO&G-related processes and issues (Perry 2012). A holistic system of a transparent mandatory risk assessment could thus be implemented by considering site-specific data including results from previous risk assessments, social perception studies, and opinions of concerned parties (Torres et al. 2016). In addition to the actual risk assessment, it is important that risk assessment reports be up to standard. However, recently reviewed environmental impact statements produced between 1998 and 2008 were found to be of poor quality concerning environmental impact prediction and project decommissioning (Anifowose et al. 2016). Therefore, a systematic and independent reviewing of these reports is recommended in order to ensure good quality.

Rahm and Riha (2012) conclude that water quality requirements can be better ensured by weighing the need for energy development and for environmental health by imposing adequate regulation. On the one hand, energy supply and water quality could be considered equally important. In this case, a flexible regime considering interests from both parties would be adequate. On the other hand, focus could be primarily on environmental and human health protection. In this case a strict precautionary principle regime would be considered as an adequate regulation. There is a need for more data in order to evaluate the risk to the water system associated with hydraulic fracturing (Gagnon et al. 2015), and even if all the risks were identified, they can never be entirely eliminated. Such a regulation would require operators to conduct research in order to address the uncertainties and knowledge gaps related to risks of UO&G operations. Risk assessment and monitoring can, moreover, be a rich source for developing environmental fate models and updating regulative and authorizations systems, which may be at the basis of UO&G-specific (legal) guidelines for (ground)water quality monitoring.

5.4   Conclusions and Way Forward

We have identified a number of uncertainties and knowledge gaps in this paper. There is a need for more chemical-based risk assessment, especially on other types of UO&G than shale gas, and in other countries than the USA. This is important for comparative reasons and ultimately for more effective regulation of UO&G operations and the ongoing updating of authorization systems. More research is also needed to verify the applicability of current environmental fate models to UO&G-related risk scenarios. Due to the lack of detailed UO&G-specific guidelines for monitoring, an effect-based monitoring may be implemented in relevant legislation to efficiently detect adverse effects to the surface water and groundwater systems. In view of the limited chronic toxicity data for UO&G-related chemicals, an update of regulation and authorization systems is required. The introduction of a system based on mandatory, location-specific, and phase-specific risk assessments for permissions is recommended, with a holistic approach to risk assessment, including site-specific data, previous risk assessment data, and social risk perception. Additionally, systematic and independent reviewing of risk assessment reports is encouraged to ensure a well-founded quality of information. There is no specific UO&G regulation on the federal US level and EU level but rather general legislation that is partly applicable to UO&G operations. More specific UO&G-tailored regulation could, however, be more effective in increasing both uniformity in practices and environmental/human health protection. Lastly, in view of the uncertainties related to UO&G risks, the precautionary principle may be given a more central role in UO&G regulation.

6 Summary

The present study identifies uncertainties and knowledge gaps of chemical risk assessment related to unconventional drillings, and proposes adaptations. A discussion is provided demonstrating that chemical risk assessment in the context of unconventional oil and gas (UO&G) activities differs from conventional chemical risk assessment, and this has implications for existing legislation. A suspect list of 1386 chemicals that might be expected in the UO&G water samples was drafted. The list can be used for LC-HRMS suspect screening. An overview of reported concentrations of substances in UO&G-related water is presented. Most information relates to shale gas operations, followed by coal-bed methane while only little is available for tight gas and conventional gas. The limited research on conventional oil and gas recovery hampers a comparative assessment of risks related to unconventional activities and those related to conventional activities. No study analyzed the whole cycle from fracturing fluid, flowback and produced water, and surface- and groundwater. In the majority of studies target screening has been used, probably missing contaminants of concern. Almost half of the organic compounds analyzed in surface water and groundwater exceed the threshold of toxicological concern values, so further risk assessment is needed and risks cannot be waived. Specific exposure scenarios towards groundwater aquifers do not exist for UO&G related activities. Human errors in various stages of the life cycle of UO&G production play an important role in the exposure. Neither at the international level nor at the US federal and the EU levels, specific regulations for UO&G related activities are in place to protect environmental and human health. UO&G activities are mostly regulated through general environmental, planning, and mining legislation.

Acknowledgments This work is part of the research program "Shale Gas & Water" with project number 859.14.001, which is financed by the Netherlands Organization for Scientific Research (NWO) and the drinking water companies Brabant Water, Oasen, and Waterleiding Maatschappij Limburg.

The authors declare that they have no conflict of interest.

References

Abualfaraj N, Gurian PL, Olson MS (2014) Characterization of Marcellus shale flowback water. Environ Eng Sci 31:514–524

ACGIH Worldwide (2013) Documentation of the TLVs and BEIs with other world wide occupational exposure values. Cincinnati

Afenyo M, Khan F, Veitch B, Yang M (2017) A probabilistic ecological risk model for Arctic marine oil spills. J Environ Chem Eng 5:1494–1503

Alawattegama SK, Kondratyuk T, Krynock R, Bricker M, Rutter JK, Bain DJ, Stolz JF (2015) Well water contamination in a rural community in southwestern Pennsylvania near unconventional shale gas extraction. J Environ Sci Health 50:516–528

Alley B, Beebe A, Rodgers J, Castle JW (2011) Chemical and physical characterization of produced waters from conventional and unconventional fossil fuel resources. Chemosphere 85:74–82

Anifowose B, Lawler DM, Van der Horst D, Chapman L (2016) A systematic quality assessment of environmental impact statements in the oil and gas industry. Sci Total Environ 572:570–585

Annevelink MPJA, Meesters JAJ, Hendriks AJ (2016) Environmental contamination due to shale gas development. Sci Total Environ 550:431–438

Arini A, Cavallin JE, Berninger JP, Marfil-Vega R, Mills M, Villeneuve DL, Basu N (2016) In vivo and in vitro neurochemical-based assessments of wastewater effluents from the Maumee River area of concern. Environ Pollut 211:9–19

Atkins JP (2013) Hydraulic fracturing in Poland: a regulatory analysis. Wash Univ Glob Stud Law Rev 12:339–355

Baars AJ, Theelen RMC, Janssen PJCM, Hesse JM, Van Apeldoorn ME, Meijerink MCM, Verdam, L, Zeilmaker M (2001) Re-evaluation of human-toxicological maximum permissible risk levels. Report no. 711701025, National Institute of Public Health and the Environment, Bilthoven, The Netherlands

Backhaus T, Faust M (2012) Predictive environmental risk assessment of chemical mixtures: a conceptual framework. Environ Sci Technol 46:2564–2573

Barbot E, Vidic NS, Gregory KB, Vidic RD (2013) Spatial and temporal correlation of water quality parameters of produced waters from Devonian-age shale following hydraulic fracturing. Environ Sci Technol 47:2562–2569

Bletsou AA, Jeon J, Hollender J, Archontaki E, Thomaidis NS (2015) Targeted and non-targeted liquid chromatography-mass spectrometric workflows for identification of transformation products of emerging pollutants in the aquatic environment. Trends Anal Chem 66:32–44

Boschee P (2014) Produced and flowback water recycling and reuse: economics, limitations, and technology. Oil Gas Facil 3:16–22

Brack W, Dulio V, Ågerstrand M et al (2017) Towards the review of the European Union water framework management of chemical contamination in European surface water resources. Sci Total Environ 576:720–737

Brady WL, Crannell JP (2012) Hydraulic fracturing regulation in the United States: the Laissez-Faire approach of the Federal Government and varying state regulations. Vermont J Environ Law 14:39–70

Brand W, de Jong CM, Van der Linden SC et al (2013) Trigger values for investigation of hormonal activity in drinking water and its sources using CALUX bioassays. Environ Int 55:109–118

Brans MC, Van den Brink KM (2014) Schaliegas in publiekrechtelijk kader – Lessen uit de praktijk van conventionele gaswinning. Milieu Recht 32:150–162

Burton TG, Rifai HS, Hildenbrand ZL, Carlton DD, Fontenot BE, Schug KA (2016) Elucidating hydraulic fracturing impacts on groundwater quality using a regional geospatial statistical modeling approach. Sci Total Environ 545:114–126

Butkovskyi A, Bruning H, Kools SA, Rijnaarts HH, Van Wezel AP (2017) Organic pollutants in shale gas flowback and produced waters: identification, potential ecological impact, and implications for treatment strategies. Environ Sci Technol 51:4740–4754

Camarillo MK, Domen JK, Stringfellow WT (2016) Physical-chemical evaluation of hydraulic fracturing chemicals in the context of produced water treatment. J Environ Manag 183:164–174

Centner TJ, O'Connell LK (2014) Unfinished business in the regulation of shale gas production in the United States. Sci Total Environ 476:359–367

Centner TJ, Petetin L (2015) Permitting program with best management practices for shale gas wells to safeguard public health. J Environ Manag 163:174–183

Chapman EC, Capo RC, Stewart BW, Kirby CS, Hammack RW, Schroeder KT, Edenborn HM (2012) Geochemical and strontium isotope characterization of produced waters from Marcellus shale natural gas extraction. Environ Sci Technol 46:3545–3553

Clark CE, Veil JA (2009) Produced water volumes and management practices in the United States. Argonne National Laboratory ANL/EVS/R-09-1. In: Prepared for the U.S. Department of

Energy, National Energy Technology Laboratory, September, p 64. http://www.ead.anl.gov/pub/dsp_detail.cfm?PubID=2437

COGC (2014) Colorado Oil and Gas Conservation Commission. Inspection./incident inquiry. http://cogcc.state.co.us/. Accessed 4 Jun 2014

Cuadrilla (2016) Cuadrilla resources ePortal. https://www.cuadrillaresourceseportal.com/

Dahm KG, Guerra KL, Xu P, Drewes JE (2011) Composite geochemical database for coalbed methane produced water quality in the Rocky mountain region. Environ Sci Technol 45:7655–7663

De Jong ER (2013) Regulating uncertain risks in an innovative society: a liability law perspective. In: Hilgendorf E, Günther JP (eds) Robotik und Recht Band I. Nomosverlag, Baden-Baden, pp 163–183

De Jong ER (2016) Voorzorgverplichtingen: over aansprakelijkheidsrechtelijke normstelling voor onzekere risico's. Dissertation, Boomjuridisch, Utrecht University

DeGraeve GM, Elder RG, Woods DC, Bergman HL (1982) Effect of naphthalene and benzene on fathead minnows and rainbow trout. Arch Environ Contam Toxicol 11:487–490

DHHS ATSDR (2010) Toxicological profile for Ethylbenzene. US Department of Health and Human Services Agency for Toxic substances and disease registry. http://www.atsdr.cdc.gov/toxprofiles/tp110.pdf

ECHA (2017) European Chemicals Agency's regulated chemicals database. https://echa.europa.eu/information-on-chemicals/registered-substances. Accessed 6 Jul 2017

EFSA (2017) European food safety authority database. https://www.efsa.europa.eu/en/data/chemical-hazards-data. Accessed 5 Apr 2017

Elliott EG, Trinh P, Ma X, Leaderer BP, Ward MH, Deziel NC (2017) Unconventional oil and gas development and risk of childhood leukemia: assessing the evidence. Sci Total Environ 576:138–147

Escher B, Leusch F (2012) Bioanalytical tools in water quality assessment. IWA Publishing, London

Escher BI, Schwarzenbach RP (2002) Mechanistic studies on baseline toxicity and uncoupling of organic compounds as a basis for modeling effective membrane concentrations in aquatic organisms. Aquat Sci 64:20–35

Escher B, Allinson M, Altenburger R et al (2013) Benchmarking organic micropollutants in wastewater, recycled water and drinking water with in vitro bioassays. Environ Sci Technol 48:1940–1956

ExxonMobile (2016) ExxonMobile chemical disclosure registry. http://www.erdgassuche-in-deutschland.de/hydraulic_fracturing/frac_massnahmen.html. Accessed 6 Nov 2016

Ferrar KJ, Michanowicz DR, Christen CL, Mulcahy N, Malone SL, Sharma RK (2013) Assessment of effluent contaminants from three facilities discharging marcellus shale wastewater to surface waters in pennsylvania. Environ Sci Technol 47:3472–3481

Ferrer I, Thurman EM (2015a) Analysis of hydraulic fracturing additives by LC/Q-TOF-MS. Anal Bioanal Chem 407:6417–6428

Ferrer I, Thurman EM (2015b) Chemical constituents and analytical approaches for hydraulic fracturing waters. Trends Environ Anal Chem 5:18–25

Fleming R, Reins L (2016) Shale gas extraction, precaution and prevention: a conversation on regulatory responses. Energy Res Soc Sci 20:131–141

Fontenot BE, Hunt LR, Hildenbrand ZL, Carlton DD Jr, Oka H, Walton JL, Osorio A, Bjorndal B, Hu QH, Schug KA (2013) An evaluation of water quality in private drinking water wells near natural gas extraction sites in the Barnett shale formation. Environ Sci Technol 47:10032–10040. https://doi.org/10.1021/es4011724

Gagnon GA, Krkosek W, Anderson L, McBean E, Mohseni M, Bazri M, Mauro I (2015) Impacts of hydraulic fracturing on water quality: a review of literature, regulatory frameworks and an analysis of information gaps. Environ Rev 24:122–131

Geraets D, Reins L (2016) The case of shale gas extraction regulation in light of CETA and TTIP: another example of the frackmentation of international law. Environ Liabil 24:16–25

Goldstein BD, Brooks BW, Cohen SD et al (2014) The role of toxicological science in meeting the challenges and opportunities of hydraulic fracturing. Toxicol Sci 139:271–283

Gordalla BC, Ewers U, Frimmel FH (2013) Hydraulic fracturing: a toxicological threat for groundwater and drinking-water? Environ Earth Sci 70:3875–3893

Gouin T, Cousins I, Mackay D (2004) Comparison of two methods for obtaining degradation half-lives. Chemosphere 56:531–535

Gramatica P (2007) Principles of QSAR models validation: internal and external. Mol Inf 26:694–107

Grant MF (2016) A critical assessment of the U.S. Environmental regulation of shale gas development – what can be learned from the U.S. experience? Edinburgh Law School working paper 2016/1

Grant CJ, Weimer AB, Marks NK et al (2015) Marcellus and mercury: assessing potential impacts of unconventional natural gas extraction on aquatic ecosystems in northwestern Pennsylvania. J Environ Sci Health 50:482–500

Gregory KB, Vidic RD, Dzombak DA (2011) Water management challenges associated with the production of shale gas by hydraulic fracturing. Elements 7:181–186

Gross SA, Avens HJ, Banducci AM, Sahmel J, Panko JM, Tvermoes BE (2013) Analysis of BTEX groundwater concentrations from surface spills associated with hydraulic fracturing operations. J Air Waste Manage Assoc 63:424–432

Harkness JS, Darrah TH, Warner NR et al (2017) The geochemistry of naturally occurring methane and saline groundwater in an area of unconventional shale gas development. Geochim Cosmochim Acta 208:302–334

Harris AE, Hopkinson L, Soeder DJ (2016) Developing monitoring plans to detect spills related to natural gas production. Environ Monit Assess 188:647

Hayes T (2009) Sampling and analysis of water streams associated with the development of Marcellus shale gas. Marcellus Shale Initiative Publications Database, 10

Hayes T, Severin BF (2012) Barnett and Appalachian Shale water management and reuse technologies. Final Report. RPSEA Contract 08122-05

Heilweil VM, Grieve PL, Hynek SA, Brantley SL, Solomon DK, Risser DW (2015) Stream measurements locate thermogenic methane fluxes in groundwater discharge in an area of shale-gas development. Environ Sci Technol 49:4057–4065

Hildenbrand ZL, Carlton Jr DD, Fontenot BE et al (2015) A comprehensive analysis of groundwater quality in the Barnett Shale region. Environ Sci Technol 49:8254–8262

Hladik ML, Focazio MJ, Engle M (2014) Discharges of produced waters from oil and gas extraction via wastewater treatment plants are sources of disinfection by-products to receiving streams. Sci Total Environ 466:1085–1093

Hoelzer K, Sumner AJ, Karatum O et al (2016) Indications of transformation products from hydraulic fracturing additives in shale-gas wastewater. Environ Sci Technol 50:8036–8048

Hogenboom AC, Van Leerdam JA, De Voogt P (2009) Accurate mass screening and identification of emerging contaminants in environmental samples by liquid chromatography-hybrid linear ion trap Orbitrap mass spectrometry. J Chromatogr A 1216:510–519

Hollander A, Schoorl M, Van de Meent D (2016) SimpleBox 4.0: improving the model while keeping it simple…. Chemosphere 148:99–107

Howarth RW, Ingraffea A, Engelder T (2011) Natural gas: should fracking stop? Nature 477:271–275

Huang PM (1980) Adsorption processes in soil. In: Hutzinger O (ed) The handbook of environmental chemistry, vol 2 Part A, Reactions and processes. Springer-Verlag, New York, pp 47–59

Hug C, Ulrich N, Schulze T, Brack W, Krauss M (2014) Identification of novel micropollutants in wastewater by a combination of suspect and nontarget screening. Environ Pollut 184:25–32

IARC (2017) Monographs on the evaluation of the carcinogenic risk of chemicals to humans. International Agency for Research on Cancer, 1972-PRESENT. http://monographs.iarc.fr/ENG/Classification/index.php. Accessed 5 Apr 2017

IEA (2017) International Energy Agency's unconventional gas production database. https://www. iea.org/ugforum/ugd/

Jackson RB, Vengosh A, Darrah TH, Warner NR, Down A, Poreda RJ, Osborn SG, Zhao K, Karr JD (2013a) Increased stray gas abundance in a subset of drinking water wells near Marcellus shale gas extraction. Proc Natl Acad Sci 110:11250–11255

Jackson RE, Gorody AW, Mayer B, Roy JW, Ryan MC, Van Stempvoort DR (2013b) Groundwater protection and unconventional gas extraction: the critical need for field-based hydrogeological research. Ground Water 51:488–510

Kahrilas GA, Blotevogel J, Stewart PS, Borch T (2014) Biocides in hydraulic fracturing fluids: a critical review of their usage, mobility, degradation, and toxicity. Environ Sci Technol 49:16–32

Kahrilas GA, Blotevogel J, Corrin ER, Borch T (2016) Downhole transformation of the hydraulic fracturing fluid biocide glutaraldehyde: implications for flowback and produced water quality. Environ Sci Technol 50:11414–11423

Kassotis CD, Tillitt DE, Davis JW, Hormann AM, Nagel SC (2013) Estrogen and androgen receptor activities of hydraulic fracturing chemicals and surface and ground water in a drilling-dense region. Endocrinology 155:897–907

Kassotis CD, Klemp KC, Vu DC et al (2015) Endocrine-disrupting activity of hydraulic fracturing chemicals and adverse health outcomes after prenatal exposure in male mice. Endocrinology 156:4458–4473

Kassotis CD, Tillitt DE, Lin CH, McElroy JA, Nagel SC (2016a) Endocrine-disrupting chemicals and oil and natural gas operations: potential environmental contamination and recommendations to assess complex environmental mixtures. Environ Health Perspect 124:256–264

Kassotis CD, Bromfield JJ, Klemp KC et al (2016b) Adverse reproductive and developmental health outcomes following prenatal exposure to a hydraulic fracturing chemical mixture in female C57Bl/6 mice. Endocrinology 157:3469–3348

Kevelam J (2015) De juridische bescherming van drinkwaterbronnen bij schaliegaswinning. Dissertation, Utrecht University

King GE (2012) Hydraulic fracturing 101: what every representative, environmentalist, regulator, reporter, investor, university researcher, neighbor and engineer should know about estimating frac risk and improving frac performance in unconventional gas and oil wells. In: Society of Petroleum Engineers Hydraulic Fracturing Technology Conference, Texas, pp 1–80

Klein W (1989) Mobility of environmental chemicals, including abiotic degradation. In: Bourdeau P, Haines JA, Klein W, Krishna Murti CR (eds) Ecotoxicology and climate: with special reference to hot and cold climates. Wiley, Chichester, pp 65–78

Kolkman A, Schriks M, Brand W et al (2013) Sample preparation for combined chemical analysis and in vitro bioassay application in water quality assessment. Environ Toxicol Pharmacol 36:1291–1303

Kondash AJ, Albright E, Vengosh A (2017) Quantity of flowback and produced waters from unconventional oil and gas exploration. Sci Total Environ 574:314–321

Krauss M, Singer H, Hollender J (2010) LC–high resolution MS in environmental analysis: from target screening to the identification of unknowns. Anal Bioanal Chem 397:943–951

Kroes R, Kleiner J, Renwick A (2005) The threshold of toxicological concern concept in risk assessment. Toxicol Sci 86:226–230

Kühne R, Ebert RU, von der Ohe PC, Ulrich N, Brack W, Schüürmann G (2013) Read-across prediction of the acute toxicity of organic compounds toward the water flea daphnia magna. Mol Inf 32:108–120

Lange T, Sauter M, Heitfeld M et al (2013) Hydraulic fracturing in unconventional gas reservoirs: risks in the geological system part 1. Environ Earth Sci 70:3839–3853

Lee Y, von Gunten U (2012) Quantitative structure – activity relationships (QSARs) for the transformation of organic micropollutants during oxidative water treatment. Water Res 46:6177–6195

Leendert V, Van Langenhove H, Demeestere K (2015) Trends in liquid chromatography coupled to high-resolution mass spectrometry for multi-residue analysis of organic micropollutants in aquatic environments. Trends Anal Chem 67:192–208

Lester Y, Ferrer I, Thurman EM, Sitterley KA, Korak JA, Aiken G, Linden KG (2015) Characterization of hydraulic fracturing flowback water in Colorado: implications for water treatment. Sci Total Environ 512:637–644

Leusch FD, Neale PA, Hebert A, Scheurer M, Schriks MC (2017) Analysis of the sensitivity of in vitro bioassays for androgenic, progestagenic, glucocorticoid, thyroid and estrogenic activity: suitability for drinking and environmental waters. Environ Int 99:120–130

Mackay D (1980) Solubility, partition coefficients, volatility, and evaporation rates. In: Hutzinger O (ed) A handbook of environmental chemistry, vol 1. Part A Reactions and Processes. Springer-Verlag, New York, pp 31–45

Mackay D, Hughes L, Powell DE, Kim J (2014) An updated Quantitative Water Air Sediment Interaction (QWASI) model for evaluating chemical fate and input parameter sensitivities in aquatic systems: application to D5 (decamethylcyclopentasiloxane) and PCB-180 in two lakes. Chemosphere 111:359–365

Maguire-Boyle SJ, Barron AR (2014) Organic compounds in produced waters from shale gas wells. Environ Sci Processes Impacts 16:2237–2248

Malaj E, Peter C, Grote M, Kühne R et al (2014) Organic chemicals jeopardize the health of freshwater ecosystems on the continental scale. Proc Natl Acad Sci 111:9549–9554

Maloney KO, Baruch-Mordo S, Patterson LA, Nicot JP, Entrekin SA, Fargione JE et al (2017) Unconventional oil and gas spills: materials, volumes, and risks to surface waters in four states of the US. Sci Total Environ 581:369–377

Maule AL, Makey CM, Benson EB, Burrows IJ, Scammell MK (2013) Disclosure of hydraulic fracturing fluid chemical additives: analysis of regulations. New Solut 23:167–187

Mellert W, Deckardt K, Kaufmann W, Van Ravenzwaay B (2007) Ethylbenzene: 4-and 13-week rat oral toxicity. Arch Toxicol 81:361–370

Meng Q (2015) Spatial analysis of environment and population at risk of natural gas fracking in the state of Pennsylvania, USA. Sci Total Environ 515:198–206

Mons MN, Heringa MB, Van Genderen J (2013) Use of the Threshold of Toxicological Concern (TTC) approach for deriving target values for drinking water contaminants. Water Res 47:1666–1678

Montgomery CT, Smith MB (2010) Hydraulic fracturing: history of an enduring technology. J Pet Technol 62:26

Müller A, Schulz W, Ruck WK, Weber WH (2011) A new approach to data evaluation in the non-target screening of organic trace substances in water analysis. Chemosphere 85:1211–1219

Murk AJ, Legler J, Van Lipzig MM et al (2002) Detection of estrogenic potency in wastewater and surface water with three in vitro bioassays. Environ Toxicol Chem 21:16–23

NAM (2016) Nederlandse aardoliematschappij chemical disclosure database. http://www.nam.nl/nl/downloads/information-fracking.html. Accessed 12 Jun 2016

Navarro HA, Price CJ, Marr MC, Myers CB, Heindel JJ (1992) Final report on the developmental toxicity of naphthalene (CAS No. 91-20-3) in New Zealand white rabbits on gestational days 6 through 19. Chemistry and Life Sciences Research Triangle Institute. NTIS Technical Report No PB-92-219831, Durham. https://www.osti.gov/scitech/biblio/6992005

Nelson J, Bishay F, Van Roodselaar A, Ikonomou M, Law FC (2007) The use of in vitro bioassays to quantify endocrine disrupting chemicals in municipal wastewater treatment plant effluents. Sci Total Environ 374:80–90

NGS (2016) Natural gas from shale hydraulic fracturing fluid and additive component transparency service. http://www.ngsfacts.org/findawell/. Accessed 6 Oct 2016

NRC (2014) National response centre. http://www.nrc.uscg.mil/. Accessed 4 Jun 2014

NTP (1990) Toxicology and carcinogenesis studies of Toluene (CAS No. 108-88-3) in F344/N Rats and B6C3F1 Mice (Inhalation Studies). National Toxicology Program Technical Report Series 371:1. https://ntp.niehs.nih.gov/ntp/htdocs/lt_rpts/tr371.pdf

O'Malley D, Karra S, Currier RP, Makedonska N, Hyman JD, Viswanathan HS (2015) Where does water go during hydraulic fracturing? Ground Water 54:488–497

OCD (2014) State of Mexico oil conservation division. Lozing search. https://wwwapps.emnrd. state.nm.us/ocd/ocdpermitting/Data/Incidents/SpillsearchResults.aspx. Accessed 4 Jun 2014

Olsson O, Weichgrebe D, Rosenwinkel KH (2013) Hydraulic fracturing wastewater in Germany: composition, treatment, concerns. Environ Earth Sci 70:3895–3906

OPPPW (2016) The polish exploration and production industry organization chemical disclosure database. http://www.opppw.pl/en/fracturing_fluid_composition/23. Accessed 6 Oct 2016

Orem W, Tatu C, Varonka M (2014) Organic substances in produced and formation water from unconventional natural gas extraction in coal and shale. Int J Coal Geol 126:20–31

Osborn SG, Vengosh A, Warner NR, Jackson RB (2011) Methane contamination of drinking water accompanying gas-well drilling and hydraulic fracturing. Proc Natl Acad Sci 108:8172–8176

PADEP (2014) Pennsylvania Department of Environmental Protection OC compliance. http://www.depreportingservices.state.pa.us/ReportServer/Pages/ReportViewer.aspx?/Oil_Gas/OG_Compliance. Accessed 4 Jun 2014

Patterson LA, Konschnik KE, Wiseman H et al (2017) Unconventional oil and gas spills: risks, mitigation priorities, and state reporting requirements. Environ Sci Technol 51:2563–2573

Pereira E, Reins L, Zhang L, Costa H, Nace J (2016) Regulatory and contractual challenges for unconventional resources: an international perspective. UEF Energy Law Rev 1:1–38

Perry SL (2012) Environmental reviews and case studies: addressing the societal costs of unconventional oil and gas exploration and production: a framework for evaluating short-term, future, and cumulative risks and uncertainties of hydrofracking. Environ Pract 14:352–365

Polishchuk K (2017) Shale gas regulation in the USA, Poland and Ukraine: managing groundwater contamination risks. Dissertation, Utrecht University

Poulsen A, Chapman H, Leusch F, Escher B (2011) Application of bioanalytical tools for water quality assessment. Urban Water Security Research Alliance Technical Report No 41, Brisbane. https://www.researchgate.net/publication/267824499_Application_of_Bioanalytical_Tools_for_Water_Quality_Assessment

Preston TM, Chesley-Preston TL (2015) Risk assessment of brine contamination to aquatic resources from energy development in glacial drift deposits: Williston Basin, USA. Sci Total Environ 508:534–545

Prpich G, Coulon F, Anthony EJ (2016) Review of the scientific evidence to support environmental risk assessment of shale gas development in the UK. Sci Total Environ 563:731–740

Rahm BG, Riha SJ (2012) Toward strategic management of shale gas development: regional, collective impacts on water resources. Environ Sci Pol 17:12–23

Randall A (2011) Risk and precaution. Cambridge University Press, Cambridge

Reemtsma T, Berger U, Arp HPH, Gallard H, Knepper TP, Neumann M, Quintana JB, Voogt PD (2016) Mind the Gap: Persistent and Mobile Organic Compounds: Water Contaminants that Slip Through. Environ Sci Technol 50:10308–10315

Reins L (2014a) European minimum principles for shale gas: preliminary insights with reference to the precautionary principle. Environ Liabil 22:16–27

Reins L (2014b) A research agenda for shale gas: challenges to a coherent regulation in the European Union. J Renew Energy Law Policy 5:167–171

Reins L (2017) Regulating shale gas: the challenge of coherent environmental and energy regulation. Edward Elgar Publishing, Cheltenham

Riedl J, Rotter S, Faetsch S, Schmitt-Jansen M, Altenburger R (2013) Proposal for applying a component-based mixture approach for ecotoxicological assessment of fracturing fluids. Environ Earth Sci 70:3907–3920

Roberson TW (2012) Environmental concerns of hydraulically fracturing a natural gas well. Utah Environ Law Rev 12:67–136

RRC (2014) Railroad commission of Texas loss report database. http://www.rrc.state.tx.us/oil-gas/compliance-enforcement/h-8/

SCHER, SCENIHR, SCCS (2012) Opinion on the toxicity and assessment of chemical mixtures. Scientific committees on health and environmental risks (SCHER), Emerging and Newly Identified Health Risks (SCENIHR) and Consumer Safety (SCCS), European Commission. http://ec.europa.eu/health/scientific_committees/environmental_risks/docs/scher_o_155.pdf

Schnoor JL (1996) Environmental modeling: fate and transport of pollutants in water, air, and soil. Wiley, New York

Scholz S, Sela E, Blaha L (2013) A European perspective on alternatives to animal testing for environmental hazard identification and risk assessment. Regul Toxicol Pharmacol 67:506–530

Schriks M, Heringa MB, Van der Kooi MM, de Voogt P, Van Wezel AP (2010a) Toxicological relevance of emerging contaminants for drinking water quality. Water Res 44:461–476

Schriks M, van Leerdam JA, Van der Linden SC, Van der Burg B, Van Wezel AP, de Voogt P (2010b) High-resolution mass spectrometric identification and quantification of glucocorticoid compounds in various wastewaters in The Netherlands. Environ Sci Technol 44:4766–4774

Schymanski EL, Singer HP, Longrée P et al (2014a) Strategies to characterize polar organic contamination in wastewater: exploring the capability of high resolution mass spectrometry. Environ Sci Technol 48:1811–1181

Schymanski EL, Jeon J, Gulde R, Fenner K, Ruff M, Singer HP, Hollender J (2014b) Identifying small molecules via high resolution mass spectrometry: communicating confidence. Environ Sci Technol 48:2097–2098

Schymanski EL, Singer HP, Slobodnik J (2015) Non-target screening with high-resolution mass spectrometry: critical review using a collaborative trial on water analysis. Anal Bioanal Chem 407:6237–6255

Shonkoff SB, Hays J, Finkel ML (2014) Environmental public health dimensions of shale and tight gas development. Environ Health Perspect 122:787

Sjerps RM, Vughs D, Van Leerdam JA, ter Laak TL, Van Wezel AP (2016) Data-driven prioritization of chemicals for various water types using suspect screening LC-HRMS. Water Res 93:254–264

Smit CE, Verbruggen EMJ (2012) Environmental risk limits for ethyl-benzene and tributyl-phosphate in water: a proposal for water quality standards in accordance with the Water Framework Directive. RIVM letter report 601714019. http://rivm.openrepository.com/rivm/handle/10029/260138

Soeder DJ (2015) Adventures in groundwater monitoring: why has it been so difficult to obtain groundwater data near shale gas wells? Environ Geosci 22:139–148

Soeder DJ, Sharma S, Pckney N, Hopkinson L, Dilmore R, Kutchko B, Stewart B, Carter K, Hakala A, Capo R (2014) An approach for assessing engineering risk from shale gas wells in the United States. Int J Coal Geol 1226:4–19

Sophocleous M (2002) Interactions between groundwater and surface water: the state of the science. Hydrogeol J 10:52–67

Stringfellow WT, Domen JK, Camarillo MK, Sandelin WL, Borglin S (2014) Physical, chemical, and biological characteristics of compounds used in hydraulic fracturing. J Hazard Mater 275:37–54

Strong LC, Gould T, Kasinkas L, Sadowsky MJ, Aksan A, Wackett LP (2013) Biodegradation in waters from hydraulic fracturing: chemistry, microbiology, and engineering. J Environ Eng 140:B4013001

Tang JY, Taulis M, Edebeli J, Leusch FD, Jagals P, Jackson GP, Escher BI (2014) Chemical and bioanalytical assessment of coal seam gas associated water. Environ Chem 12:267–285

ter Heege JH, Griffioen J, Schavemaker YA, Boxem TAP (2014) Inventarisatie van technologieën en ontwikkelingen voor het verminderen van (rest) risico's bij schaliegaswinning, No. TNO-2014-R10919. TNO. https://repository.tudelft.nl/view/tno/uuid:d3f72f6d-c669-4e8a-b4aa-fa87e399bf00/

Thacker JB, Carlton DD, Hildenbrand ZL, Kadjo AF, Schug KA (2015) Chemical analysis of wastewater from unconventional drilling operations. Water 7:1568–1579

Theodori GL, Luloff AE, Willits FK, Burnett DB (2014) Hydraulic fracturing and the management, disposal, and reuse of frac flowback waters: views from the public in the Marcellus Shale. Energy Res Soc Sci 2:66–74

Torres L, Yadav OP, Khan E (2016) A review on risk assessment techniques for hydraulic fracturing water and produced water management implemented in onshore unconventional oil and gas production. Sci Total Environ 539:478–493

US EPA (2015) Assessment of the potential impacts of hydraulic fracturing for oil and gas on drinking water resources. Office of Research and Development, US Environmental Protection Agency, Washington

US EPA (2016) Hydraulic fracturing for oil and gas: impacts from the hydraulic fracturing water cycle on drinking water resources in the United States. Office of Research and Development, U.S. Environmental Protection Agency, Washington

US EPA Ecotox (2017) U.S. Environmental Protection Agency's Ecotox database. https://cfpub. epa.gov/ecotox. Accessed 6 Apr 2017

US EPA IRIS (2017) U.S. Environmental Protection Agency's integrated risk information system. https://www.epa.gov/iris. Accessed 6 Apr 2017

US Fracfocus (2016) https://fracfocus.org/chemical-use/chemicals-public-disclosure. Accessed 5 Feb 2016

Van Herwijnen R, Vos JH (2009) Environmental risk limits for benzene, C10-13 alkyl derivs (LAB). RIVM Report No 601782016. http://www.rivm.nl/dsresource?objectid=a837022c-94e9-481e-8d32b6cd3104c826&type=org&disposition=inline

Van Wezel AP, Ter Laak TL, Fischer A, Bäuerlein PS, Munthe J, Posthuma L (2017) Mitigation options for chemicals of emerging concern in surface waters; operationalising solutions-focused risk assessment. Environ Sci Water Res Technol 3:403–414

Venkatesan AK, Halden RU (2015) Effective strategies for monitoring and regulating chemical mixtures and contaminants sharing pathways of toxicity. Int J Environ Res Public Health 12:10549–10557

Vidic RD, Brantley SL, Vandenbossche JM, Yoxtheimer D, Abad JD (2013) Impact of shale gas development on regional water quality. Science 340:6134

Vos A (2014) Shale gas extraction: in line with the general (environmental) principles of Union and Dutch law? Dissertation, Utrecht University

Wachinger G, Renn O, Begg C, Kuhlicke C (2013) The risk perception paradox implications for governance and communication of natural hazards. Risk Anal 33:1049–1065

Warner NR, Christie CA, Jackson RB, Vengosh A (2013a) Impacts of shale gas wastewater disposal on water quality in western Pennsylvania. Environ Sci Technol 47:11849–11857

Warner NR, Kresse TM, Hays PD, Down A, Karr JD, Jackson RB, Vengosh A (2013b) Geochemical and isotopic variations in shallow groundwater in areas of the Fayetteville Shale development, north-central Arkansas. Appl Geochem 35:207–220

Webb E, Bushkin-Bedient S, Cheng A, Kassotis CD, Balise V, Nagel SC (2014) Developmental and reproductive effects of chemicals associated with unconventional oil and natural gas operations. Rev Environ Health 29:307–318

Werner AK, Vink S, Watt K, Jagals P (2015) Environmental health impacts of unconventional natural gas development: a review of the current strength of evidence. Sci Total Environ 505:1127–1141

Westerhoff P, Yoon Y, Snyder S, Wert E (2005) Fate of endocrine-disruptor, pharmaceutical, and personal care product chemicals during simulated drinking water treatment processes. Environ Sci Technol 39:6649–6663

Yost EE, Stanek J, DeWoskin RS, Burgoon LD (2016) Estimating the potential toxicity of chemicals associated with hydraulic fracturing operations using quantitative structure-activity relationship modeling. Environ Sci Technol 50:7732–7742

Zhang M, Wang Y, Wang Q, Yang J, Yang D, Liu J, Li J (2010) Involvement of mitochondria-mediated apoptosis in ethylbenzene-induced renal toxicity in rat. Toxicol Sci 115:295–303

Ziemkiewicz PF, Quaranta JD, Darnell A, Wise R (2014) Exposure pathways related to shale gas development and procedures for reducing environmental and public risk. J Nat Gas Sci Eng 16:77–84

Zoveidavianpoor M, Samsuri A, Shadizadeh SR (2012) Overview of environmental management by drill cutting re-injection through hydraulic fracturing in upstream oil and gas industry. In: Curkovic S (ed) Sustainable development – authoritative and leading edge content for environmental management. InTech. https://www.intechopen.com/books/sustainable-develop ment-authoritative-and-leading-edge-content-for-environmental-management/overview-of-environmental-management-by-drill-cutting-re-injection-through-hydraulic-fracturing-in-u

A Review of the Chemistry, Pesticide Use, and Environmental Fate of Sulfur Dioxide, as Used in California

Kelsey Craig

Contents

Abbreviations

ATSDR   Agency for Toxic Substances and Disease Registry
CARB   California Air Resources Board
DPR   California Department of Pesticide Regulation
K_{oc} Soil adsorption coefficient
K_{ow} Octanol-water partition coefficient
NAAQS   National Ambient Air Quality Standards

K. Craig (✉)
California Department of Pesticide Regulation, Environmental Monitoring Branch:
Air Program, Sacramento, CA, USA
e-mail: kelsey.craig@cdpr.ca.gov

© Springer International Publishing AG 2018
P. de Voogt (ed.), *Reviews of Environmental Contamination and Toxicology*
Volume 246, Reviews of Environmental Contamination and Toxicology 246,
DOI 10.1007/398_2018_11

NAMS   National Air Monitoring Stations
PUR   Pesticide Use Reporting
SLAMS   State and Local Air Monitoring Stations
USDA   United States Department of Agriculture
USEPA   United States Environmental Protection Agency

1 Introduction

Sulfur dioxide (SO_2) is used as a fungicide for the treatment of postharvest grape products, typically for the prevention of gray mold disease (*Botrytis cinerea*) in cold-storage warehouses and trucks, vans, trailers, or train cars (US EPA 2014a). In wineries, SO_2 is used to sanitize wine barrels, corks, and tanks and to prevent mold or bacterial growth (DPR 2009). Uses of SO_2 as an antioxidant and to inhibit wild yeast growth during winemaking are considered non-pesticide uses (DPR 2016a).

Other products can produce SO_2, either as an active ingredient to control pests or as a pesticide use by-product. For example, pads containing anhydrous sodium metabisulfite are used to prevent growth of *B. cinerea* during shipment of grapes and release 1–5 ppm SO_2 upon absorption of ambient moisture (US EPA 2013). Rodenticide smoke bombs produce SO_2 upon combustion of elemental sulfur (USDA 2011). Additionally, agricultural applications of sulfur release SO_2 into the environment as a product of oxidation (Griffith et al. 2015). Sulfur is the largest-volume pesticide in use globally (Griffith et al. 2015), with early historical uses as a disinfectant dating back to ancient Greece and Rome (Nriagu 1978).

In California, SO_2 is also used in combination with carbon dioxide to treat stored postharvest grapes for black widow spider (*Latrodectus hesperus*) under the Federal Insecticide, Fungicide, and Rodenticide Act (FIFRA) Section 24(c) special local need (SLN) registration process (DPR 2011a). DPR has authority to grant SLN registration applications from registrants or third-party applicants if all of the following apply: the SLN cannot be alleviated by a currently registered product; the active ingredients are federally registered by the US EPA; food residue tolerances or exemptions exist; and the product use has not previously been denied, suspended, or cancelled (DPR 2011b).

Both natural and anthropogenic sources contribute to ambient atmospheric concentrations of SO_2 in the United States (US EPA 2008a). Anthropogenic SO_2 emissions are mainly from the combustion of fossil fuels for power production (73%) and other industrial activities (20%) (US EPA 2013). Natural sources of SO_2 include volcanoes and wildfires (US EPA 2010). Use of coal-fired power plants results in the production of substantially greater SO_2 emissions from anthropogenic sources compared to natural sources (Smith et al. 2011), despite recent considerable anthropogenic emission reductions (US EPA 2014b). For example, there was a 73% reduction of SO_2 emissions in the United States from 1990 to 2011 (US EPA 2014b). Agricultural sources including agricultural pesticide applications, livestock waste, and agricultural field burning contributed less than 1% of the total SO_2 emissions tracked by the US EPA National Emissions Inventory in 2011 (US EPA 2015).

Under the Clean Air Act, US EPA sets the National Ambient Air Quality Standards (NAAQS), which include SO_2, as an indicator pollutant of sulfur oxides (US EPA 2008b). Primary NAAQS have been established for the protection of human health, while secondary NAAQS are protective of the environmental and public welfare impacts of sulfur oxides, such as acid rain (US EPA 2010).

This report reviews the relevant literature addressing the chemical properties, environmental fate, and pesticide uses of SO_2 in California. Special attention is given to understanding the atmospheric, aquatic, and terrestrial fate of SO_2 in the environment.

2 Physical and Chemical Properties

Table 1 lists the physical and chemical properties of SO_2. SO_2 is a colorless, non-flammable, volatile gas at room temperature and standard atmospheric pressure with a strong, pungent odor (Gammon et al. 2010). SO_2 is water soluble and forms hydrated SO_2 ($SO_2 \cdot H_2O$), bisulfite ions (HSO_3^-), and sulfite ions (SO_3^{2-}) upon dissolution in water (Eq. 1). Due to the decreased solubility of SO_2 as it reacts with water, the effective Henry's law constant is greatly decreased with decreasing pH (Seinfeld and Pandis 2016).

Chemical Equation for the Dissolution and Subsequent Oxidation of SO_2 in Water
Dissolution of SO_2 and formation of hydrated SO_2, bisulfite, and sulfite ions (Seinfeld and Pandis 2016).

$$SO_{2(g)} + H_2O_{(l)} \rightleftharpoons SO_2 \cdot H_2O \rightleftharpoons HSO_3^-{}_{(aq)} + H^+{}_{(aq)} \rightleftharpoons SO_3^{2-}{}_{(aq)} + 2H^+{}_{(aq)} \quad (1)$$
$$pK_{s1} = 1.9 \qquad\qquad pK_{s2} = 7.2$$

3 Use Profile

SO_2 has been registered by the US EPA for use as a pesticide in the United States since 1988 (US EPA 2016c). US EPA is scheduled to complete a registration review of "Inorganic Sulfites (Sulfur Dioxide)" in 2019 (US EPA 2014a). Prior to distribution, sale, or use in California, pesticides must also be registered with DPR (DPR 2017a). Two products containing the listed active ingredient SO_2 are currently registered in California: "The Fruit Doctor" manufactured by Snowden Enterprises and "Airgas Sulfur Dioxide" manufactured by AirGas, USA LLC (DPR 2017b). (See Appendix, Table 6, for basic information on product formulations and uses.)

Table 1 Physical and chemical properties of sulfur dioxide

Property (unit)[a]	Value	Property (unit)[b]	Value
Chemical name	Sulfur dioxide	Vapor density (g/L)	2.927
Synonyms[c]	Sulfur (VI) oxide Sulfur superoxide Sulfurous acid anhy- dride Sulfurous anhydride	Liquid density (g/L)	1.434
Empirical formula	SO_2	Specific gravity	2.26
CAS registry number	7446-09-5	Relative vapor density (air = 1)	2.25
Physical state	Gas	Relative liquid density $-10°C$ (water = 1)	1.4
Color	Colorless	Log K_{ow}[d, e]	No data
Molecular weight (g/mol)	64.07	Log K_{oc}[d]	No data
Melting point[f] (°C)	-75.5	Henry's law constant[g] (mol/m^3Pa)	1.3×10^{-2}
Boiling point[f] (°C)	-10	Vapor pressure 20°C (mmHg)	3,000
Odor threshold[h] (ppm)	0.45	Solubility in water 0°C (g/100 mL)	22.8
Conversion factor[i] (gas, 25°C, 101.3 kPa)	1 ppm = 2.62 mg/m³	Solubility in water 20°C (g/100 mL)	11.3

[a]US EPA (2013), unless otherwise noted
[b]US EPA (2007), unless otherwise noted
[c]USDA (2011)
[d]ATSDR (1998) and Griffith et al. (2015)
[e]Estimated value, -2.20 (US EPA 2012)
[f]Ashar (2016)
[g]Sander et al. (2011), as cited by Sander (2015)
[h]Ruth (1986)
[i]Boubel et al. (1994)

Pesticide products containing SO_2 for fumigations or sterilizations are available in cylinders of compressed liquid SO_2 that converts to a gas upon release (US EPA 2007). Six pesticide products with SO_2 as the listed active ingredient were reportedly used in California during 2010–2015 (DPR 2017d). Use of the product "The Fruit Doctor" exceeded that of all other SO_2 products during this period (Fig. 1).

The recommended application rate for fungicidal use of SO_2 on postharvest grapes is up to 1% gas concentration (by volume of the fumigated space) with up to 20 treatments allowable in 7–10-day intervals, depending on the variety of grape (Gammon et al. 2010; Snowden Enterprises 2016; US EPA 2013). For extended storage, initial gassing should occur on the day of harvest and continue at a frequency of three times per week at a lower concentration (Snowden Enterprises 2016; US EPA 2013). (See Appendix, Table 7 for SO_2 application rates and postharvest grape fumigation requirements.)

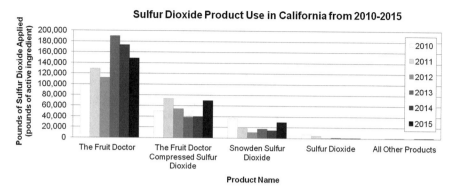

Fig. 1 Total reported annual pesticide use of sulfur dioxide (pounds of active ingredient) in California from January 1, 2010, to December 31, 2015, summarized by-product

Total utilization is a technique in which SO_2 is circulated within the fumigation chamber until almost completely absorbed into the treated commodity, resulting in low levels of SO_2 (typically less than 30 ppm) released during aeration (US EPA 2007). *B. cinerea* treatments almost exclusively employ total utilization. However, *L. hesperus* treatments (up to 10,000 ppm) do not use total utilization and may result in higher release concentrations during aeration (US EPA 2007). Therefore, US EPA established product label revision requirements including maximum release concentrations of 30 ppm for warehouse treatments and 2 ppm for truck/trailer treatments (US EPA 2007).

DPR maintains an extensive database of Pesticide Use Reporting (PUR) records for applications of pesticides within California (DPR 2017c). The level of information contained in each record depends on whether the pesticide application is considered agricultural or nonagricultural under state regulations (DPR 2017c). For example, agricultural pesticides are reported by application date and location within the Public Land Survey System, which constrains spatial resolution of agricultural pesticide applications to approximately 1 mi^2 (Craig 2017). However, nonagricultural pesticide uses (e.g., commodity fumigation of stored grapes) are reported in monthly countywide summaries. Therefore, the spatial resolution of nonagricultural pesticide application records is restricted to the county level. DPR staff queried the PUR database on August 2, 2017 to identify sources of SO_2 reported in California from January 1, 2010, to December 31, 2015 (DPR 2017d).

Figure 2 displays reported use of SO_2 as a map of the cumulative total pounds of SO_2 applied (pounds of active ingredient) in California counties from January 1, 2010, to December 31, 2015, according to the PUR database (DPR 2017d). For the top ten counties with the highest use of SO_2, the total annual pounds applied during 2010–2015 are displayed as a stacked horizontal bar graph (Fig. 3) and are tabulated for reference (Appendix, Table 8). The statewide total pounds of SO_2 applied as a pesticide ranged annually from approximately 120,000 to 250,000 pounds from 2000 to 2015, as shown in Fig. 4. The cyclic temporal trends in SO_2 use shown in Fig. 4 may be influenced by winery operations. For example, wine

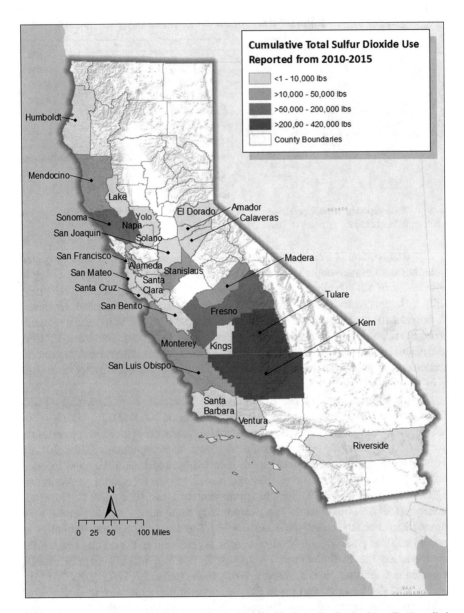

Fig. 2 The cumulative total reported pounds of sulfur dioxide (pounds of active ingredient) applied as a pesticide in California counties from January 1, 2010, to December 31, 2015, summarized by total pounds applied

barrels may be emptied and treated with SO₂ in 9–24 month cycles, depending on the wine grape variety (A. Craig, personal communication, May 1, 2017). In California, wines are aged for an average of 2 years (Bombrun and Sumner 2003). The top three highest use years occurred during more recent years (2011–2015), indicating a

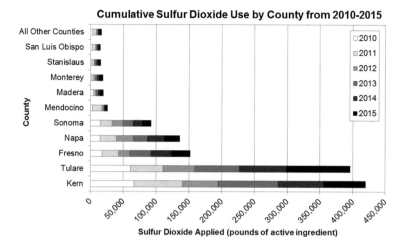

Fig. 3 Cumulative total reported pesticide use of sulfur dioxide (pounds of active ingredient) in California from January 1, 2010, to December 31, 2015, summarized by county and year

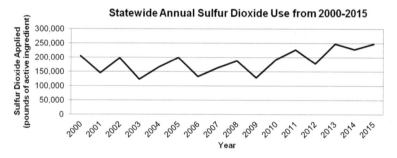

Fig. 4 Total annual pesticide use of sulfur dioxide (pounds of active ingredient) reported in California from January 1, 2000, to December 31, 2015

potential recent trend of increasing use of SO_2 as a pesticide in California. During this period, there was also an increase (8%) in reported California grape acreage due to increased acreage of table (29%) and wine (12%) grapes, although raisin (-11%) grape acreage decreased (CDFA 2014, 2015, 2016; DPR 2016b).

Figure 5 displays the average statewide total monthly use of SO_2 in California from 2010 to 2015 and shows that statewide use of SO_2 was highest during July–November. Figure 6 shows that reported SO_2 use from 2010 to 2015 in California was mainly for fumigations (96.04%). PUR records that indicated SO_2 treatments on grapes (53.02%), other fumigation (25.78%), commodity fumigation (12.97%), wine grapes (0.11%), or commercial, institutional, and industrial areas (0.01%) were considered to be reported uses for treatments of postharvest grape products or winery equipment sterilizations and were combined as "fumigations." Records without an indicated crop name (4.16%) were also considered to be fumigations.

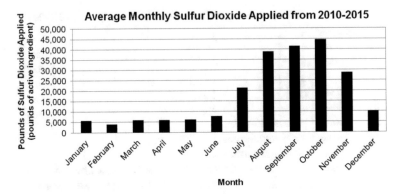

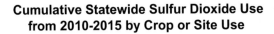

Fig. 5 Average reported monthly pesticide use of sulfur dioxide (pounds of active ingredient) in California from January 1, 2010, to December 31, 2015

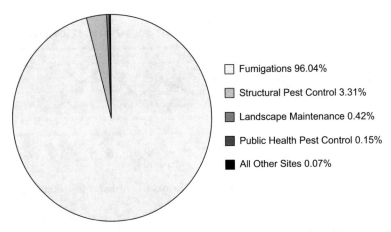

Fig. 6 Cumulative total reported pesticide use of sulfur dioxide (pounds of active ingredient) reported in California from January 1, 2010, to December 31, 2015, summarized by site

Other site uses contributed approximately 4% of reported statewide SO_2 use from 2010 to 2015 (Appendix, Table 9).

4　Environmental Fate and Degradation

SO_2 is a highly water-soluble gas (Gammon et al. 2010) that, due to a high vapor pressure (Table 1), will tend to partition in the atmosphere (US EPA 2007). Atmospheric SO_2 contributes to visibility impairment and is a major precursor to fine particulate matter (PM$_{2.5}$), a pollutant of concern for environmental and public health (US EPA 2014b). Unless removed by wet or dry deposition, atmospheric sulfur-containing compounds may travel hundreds to thousands of kilometers from their source (Smith et al. 2001; Seinfeld and Pandis 2016). Sulfur-containing compounds can acidify precipitation and soils, damaging ecosystems, property, and crops (US EPA 2014b; USDA 2011). SO$_2$ used as a pesticide is expected to enter the sulfur pool in the biosphere, participating in biogeochemical cycling processes in sulfur reaction pathways as an essential element between the environment and living organisms (Griffith et al. 2015; Moss 1978). However, anthropogenic SO$_2$ emissions have substantial influences on sulfur cycle equilibria (Faloona 2009; Moss 1978).

The following bulleted list summarizes the role of SO$_2$ in the global sulfur cycle, as illustrated in Fig. 7:

1. SO$_2$ used as a pesticide is expected to combine with other anthropogenic SO$_2$ emissions (mainly from fossil fuel use) and natural SO$_2$ emissions (mainly from volcanic eruptions) in the atmosphere (Griffith et al. 2015; US EPA 2008a). SO$_2$ is removed from the atmosphere largely by oxidation to sulfur oxides (Faloona 2009; Seinfeld and Pandis 2016).

2. Atmospheric SO$_2$ may be deposited directly to soils or vegetation in the process of dry deposition (Al-Jahdali and Bisher 2008; US EPA 2008a).

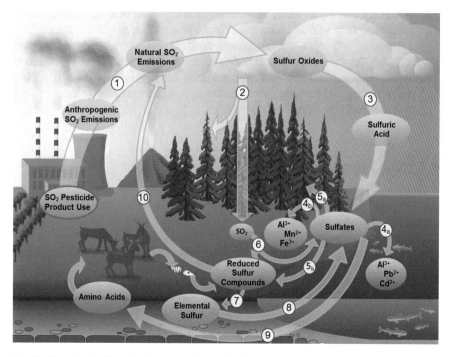

Fig. 7 The role of sulfur dioxide in the sulfur cycle (Adapted from Encyclopedia Britannica 2008)

3. Wet deposition occurs when atmospheric reactions oxidize SO_2 to produce sulfuric acid (H_2SO_4), which may deposit sulfates onto soil or surface waters as acid rain (US EPA 2014c).
4. Heavy metals such as aluminum, manganese, iron, lead, or cadmium have a greater affinity for protons than soil cations and may therefore be leached from soil in acidic conditions, which may harm (a) aquatic and (b) terrestrial organisms (Griffith et al. 2015; USDA 2011; US EPA 2016a).
5. Sulfates in the soil may be (a) absorbed by plant roots (Havlin et al. 2005; USDA 2011) or may be (b) reduced to hydrogen sulfide (H_2S) by plants or soil microbes (Moss 1978; Strawn et al. 2015).
6. Depending on the soil microbial community present, dry deposition of SO_2 to soils may result in the production of sulfates (oxidation) or reduced sulfur compounds (reduction) in soils during biodegradation (Strawn et al. 2015).
7. Soil microbial processes can also oxidize H_2S to sulfur (Havlin et al. 2005). H_2S may also volatilize from anaerobic soils and oxidize more rapidly in the atmosphere (Moss 1978).
8. Sulfur may be oxidized by aerobic soil microbes to produce sulfates (Havlin et al. 2005).
9. Sulfates are available for absorption by plant roots and are converted to amino acids, which become proteins within plant tissues (Havlin et al. 2005). Upon decomposition, anaerobic microbial degradation of amino acids may produce H_2S (Dämmgen et al. 1998).
10. Eventually, soil sulfides may be oxidized by combustion of fossil fuels or during volcanic activity to release sulfur compounds including SO_2 to the atmosphere (Encyclopedia Britannica 2008; Dämmgen et al. 1998). Other sources of natural SO_2 emissions include oxidation of reduced sulfur species (e.g., H_2S, dimethyl sulfide, carbon disulfide, carbonyl sulfide, methyl mercaptan, and dimethyl disulfide) largely produced by organisms in pelagic, coastal, estuary, or marsh environments (US EPA 2008c).

4.1 Environmental Fate and Degradation in Air

Removal of SO_2 from the atmosphere may result from oxidation, wet or dry deposition, aqueous dissolution, or absorption by soil or plant surfaces (Alberta Environment 2003; ATSDR 1998). Factors such as humidity, particulate matter composition, aerosol pH, and concentrations of reactant species influence atmospheric removal rates (Alberta Environment 2003; Huang et al. 2016; Liang and Jacobson 1999; Seinfeld and Pandis 2016). The residence time of SO_2 in the atmosphere is on the order of days (Griffith et al. 2015; US EPA 2008a) and depends upon altitude, location, and meteorological conditions, with shorter atmospheric lifetimes closer to the Earth's surface (US EPA 2008a).

Oxidation is the main removal process of SO_2 from the atmosphere (Faloona 2009; Seinfeld and Pandis 2016). Rates of atmospheric SO_2 oxidation reactions are influenced by photochemistry and temperature, such that the highest rates occur during daytime temperatures and summer conditions, due to higher concentrations of oxidants produced in photochemical reactions (Finlayson-Pitts and Pitts 1986). Oxidation of SO_2 increases the sulfur oxidation state from S(IV) to S(VI), producing bisulfate (HSO_4^-) and sulfate (SO_4^{2-}) ions (US EPA 2008a). Although there is considerable regional variation, oxidation of anthropogenic SO_2 emissions contributes 72% of sulfate aerosols globally (Forster et al. 2007; Blanco et al. 2014). Table 2 outlines various atmospheric SO_2 oxidation pathways.

Oxidation of atmospheric SO_2 involves homogeneous gas-phase reactions in air, homogeneous aqueous-phase reactions in liquid droplets, heterogeneous gas-solid reactions of gaseous SO_2 on particle surfaces, or combinations of each (Alberta Environment 2003). Atmospheric oxidation of SO_2 produces sulfur trioxide (SO_3) and sulfates (ATSDR 1998; US EPA 2008a) and has been reported to result mainly from aqueous-phase reactions (Hoyle et al. 2016). Over 70% of the atmospheric oxidation of SO_2 is estimated to occur in the aqueous-phase (Langner and Rodhe 1991), and approximately 20% of the atmospheric oxidation of SO_2 is thought to occur in the gas-phase (US EPA 2008a). However, oxidation reaction rates are influenced by meteorological and other local conditions, including reactant concentrations (US EPA 2008a).

4.1.1   Oxidation: Homogeneous Aqueous-Phase

Upon aqueous dissolution, SO_2 reacts with water to form products including HSO_3^- and SO_3^{2-} ions (US EPA 2008a). The subsequent oxidation of S(IV) species produces S(VI) species such as HSO_4^- and SO_4^{2-} ions (US EPA 2008a). The atmospheric oxidation of SO_2 is dominated by aqueous-phase reactions (Faloona 2009; Langner and Rodhe 1991; Hoyle et al. 2016). Aqueous-phase oxidation of SO_2 to sulfate is mainly due to reaction with hydrogen peroxide (H_2O_2), O_3, ˙OH, or transition metal ion catalysts (e.g., iron, manganese, or copper), and in-cloud oxidation is most often reported to be dominated by H_2O_2 (Harris et al. 2014; Hoyle et al. 2016; Seinfeld and Pandis 2016; US EPA 2008a). Equation (2) shows reaction of H_2O_2 with HSO_3^- (resulting from dissolution of SO_2 and subsequent production of HSO_3^- and SO_3^{2-}, as shown in Eq. 1).

Aqueous-Phase Oxidation Reaction of Bisulfite Ions (HSO_3^-) and Hydrogen Peroxide (H_2O_2)
Homogeneous aqueous-phase oxidation of HSO_3^- from the dissolution of sulfur dioxide in aerosols or cloud droplets in a reaction with H_2O_2 (US EPA 2008a).

$$HSO_3^-{}_{(aq)} + H_2O_{2(aq)} + H^+{}_{(aq)} \rightleftharpoons SO_4^{2-}{}_{(aq)} + H_2O_{(l)} + 2H^+{}_{(aq)} \qquad (2)$$

Liang and Jacobson (1999) reported that atmospheric oxidation of SO_2 by H_2O_2, O_3, methyl hydroperoxide, and ˙OH are sensitive to environmental conditions (i.e., pH, temperature, sunlight, and liquid water content). Table 3 shows factors

Table 2 Sulfur dioxide atmospheric oxidation reactions

Reaction phase	Mechanism	Net reaction	Reactants or catalysts	Factors influencing reaction rate
Homogeneous gas-phase	Direct photooxidation (in air)	$SO_2 \rightarrow H_2SO_4$	Light, O_2, and H_2O	Sunlight intensity, concentration of SO_2
	Indirect photooxidation (in air)	$SO_2 \rightarrow H_2SO_4$	H_2O, smog, photochemically generated reactive intermediates: $^{\cdot}OH$, $HO_2^{\cdot}$, $RO^{\cdot}$, NO_x	Concentrations of $^{\cdot}OH$, $HO_2^{\cdot}$, $RO^{\cdot}$, organic oxidants, and SO_2
	Oxidation (in air)	$SO_2 \rightarrow H_2SO_4$	Thermally generated reactive intermediates: O_3, alkenes	Concentrations of alkenes[a]
Homogeneous aqueous-phase	Oxidation (in liquid droplets and on moist surfaces)	$NH_3 + H_2SO_3 \rightarrow NH_4^+ + SO_4^{2-}$ $SO_2 \rightarrow SO_4^{2-}$	H_2O_2, O_3, $^{\cdot}OH$, SO_5^{-}, HSO_5^{-}, SO_4^{-}, PAN, CH_3COOH, $CH_3C(O)OOH$, $HO_2^{\cdot}$, NO_3, NO_2, N(III), HCHO, Cl_2^{-}	Concentration of ammonia pH, ionic strength[b, c], temperature[b]
	Catalyzed oxidation (in liquid droplets and on moist surfaces)	$SO_2 \rightarrow SO_4^{2-}$	O_2, H_2O, metal ions	Concentration of metal ions or salts (iron, vanadium, manganese)
Heterogeneous gas-solid phase	Catalyzed oxidation (on dry surfaces)	$SO_2 \rightarrow H_2SO_4$	O_2, H_2O, metal ions, particulate carbon	Carbon particle concentration (surface area)

Wilson (1978), as cited by Alberta Environment (2003), unless otherwise noted
[a]Eggleton and Cox (1978)
[b]Liang and Jacobson (1999)
[c]Seinfeld and Pandis (2016) and Bunce (1994)

Table 3 Dominant oxidants of sulfur dioxide in liquid droplets

pH range [a]	Water content range (g H_2O m^{-3})	Dominant oxidant
0–5	(3×10^{-4})–9	H_2O_2
5–6	(3×10^{-4})–9	H_2O_2 (if H_2O_2 depleted, ˙OH may be significant)
4–8	1–9	O_3
6–8	(3×10^{-4})–9	O_3

Liang and Jacobson (1999)
[a]Held at a constant pH value

influencing photochemical oxidation of SO_2 in tropospheric aerosols investigated using a gas-aqueous photochemical box model. H_2O_2 was found to be the most important oxidant in the aqueous-phase, except in aerosols of high water content and initial pH. Aqueous-phase oxidation was reported to be more important in winter vs. summer conditions (Liang and Jacobson 1999). In clouds, the rate of aqueous-phase oxidation of SO_2 by H_2O_2 has been estimated to be approximately 10% per min, given an H_2O_2 concentration of 1 ppb (Seinfeld and Pandis 2016).

Harris et al. (2014) examined sulfate sources in an air parcel traveling through an orographic cloud and reported that sulfate production in cloud droplets depends on both time of day and particle size. Oxidation of H_2O_2 was dominant in larger aerosols, and oxidation was enhanced by higher concentrations of H_2O_2 during the daytime. The authors suggested that due to the self-limiting nature of O_3 reactions (as solution pH and reaction rate decreases with SO_2 oxidation) and lower pH dependence of transition metal ion-catalyzed aqueous-phase oxidation, the major SO_2 removal processes in clouds may ultimately depend upon a droplet-size sensitive process of activation and supersaturation as well as transition metal ion concentration and composition. Due to the exhaustion of H_2O_2 during rapid oxidation of SO_2, transition metal ion catalysis was found to result in the largest amount of SO_2 oxidation in the cloud examined (Harris et al. 2014). Concentrations of other reactants such as ammonium ions (NH_4^+) may also influence rates of dissolution and oxidation of SO_2 by increasing the solution pH (US EPA 2008a).

4.1.2 Oxidation: Homogeneous Gas-Phase

Homogeneous gas-phase reactions involve either direct photochemical oxidation of electronically excited SO_2 initiated by solar energy, indirect oxidation by photochemically generated reactive intermediates, or oxidation by thermally generated reactive intermediates (Alberta Environment 2003).

Direct photolytic degradation of SO_2 requires solar irradiation of wavelengths that do not reach the troposphere (Dämmgen et al. 1998). However, direct photochemical oxidation may occur when tropospheric SO_2 molecules electronically excited by solar irradiation react with SO_2 or O_2 to form SO_3 (ATSDR 1998; Dämmgen et al. 1998). Direct photochemical oxidation is not considered an important degradation pathway due to low atmospheric concentrations of excited triplet SO_2, which is quenched by O_2 (Cox 1973).

Photochemically generated compounds including hydroxyl radicals ($^{\cdot}OH$), perhydroxyl radicals ($HO_2^{\cdot}$), or hydrocarbon radicals can oxidize SO_2 to $HSO_3^{\cdot}$ and SO_3 in homogenous gas-phase reactions (Boubel et al. 1994). The reactive gas intermediates are rapidly hydrolyzed with atmospheric moisture to form H_2SO_4 (Boubel et al. 1994). Oxidation of SO_2 by photochemically generated species occurs most often from reactions with $^{\cdot}OH$ (Alberta Environment 2003; ATSDR 1998). Equation (3) shows the reaction of SO_2 with $^{\cdot}OH$ to produce H_2SO_4. Gaseous H_2SO_4 is extremely water soluble and has a very low vapor pressure; therefore, participation of H_2SO_4 in the nucleation of new sulfate aerosols and the rapid transfer of H_2SO_4 to aerosol particles and cloud droplets may contribute to acid rain (US EPA 2008a; Seinfeld and Pandis 2016). Acid rain is the deposition of sulfuric and nitric acids from the atmosphere in a mixture of wet and dry removal processes (USDA 2011).

Photooxidation reactions involving species such as atomic oxygen (O), ozone (O_3), HO_2, organic radicals, and the Criegee biradical (RCHOOH) are negligible compared to reaction with $^{\cdot}OH$, either due to slow reaction rate with SO_2 or low atmospheric concentration of the reactants (Alberta Environment 2003). Table 4 describes the homogeneous gas-phase oxidation of SO_2 by various oxidants and clearly shows $^{\cdot}OH$ to be the dominant oxidizing species in the gas-phase (Harrison 2001). Although the contribution of stabilized Criegee intermediates (SCI) to atmospheric SO_2 oxidation may be enhanced in certain conditions (Sarwar et al. 2014), the influence of SCI on SO_2 oxidation has been estimated to be 13% of the diurnal loss rate of SO_2 to $^{\cdot}OH$ using rate constants representing upper limits (Newland et al. 2015).

Atmospheric Oxidation of Sulfur Dioxide by the Hydroxyl Radical
Atmospheric oxidation of sulfur dioxide in a homogeneous gas-phase reaction with the hydroxyl radical, where M is an atmospheric component which stabilizes the reaction product, such as N_2 or O_2 (US EPA 2008a).

$$\begin{aligned}
SO_2 + {}^{\cdot}OH + M &\rightarrow HSO_3^{\cdot} + M \\
HSO_3^{\cdot} + O_2 &\rightarrow SO_3 + HO_2^{\cdot} \\
SO_3 + H_2O &\rightarrow H_2SO_4
\end{aligned} \tag{3}$$

Oxidation of SO_2 may also be initiated by thermally generated reactive compounds including N_2O_5, NO_3, and alkenes (Alberta Environment 2003). Eggleton and Cox (1978) reported that N_2O_5 and NO_3 oxidation of SO_2 was negligible under laboratory conditions and that oxidation in the presence of alkenes and ozone was substantial only at high concentrations of alkenes. These authors also suggested that oxidation by thermally generated reactive species is likely to be important only in urban areas with high concentrations of alkenes.

Table 4 Homogeneous gas-phase oxidation of sulfur dioxide

Oxidant	Concentration[a] (cm^{-3})	Rate constant ($cm^3\ mol^{-1}\ s^{-1}$)	Loss of SO_2 (% h^{-1})
$\cdot OH$	5×10^6	9×10^{-13}	1.6
Criegee biradical	1×10^6	7×10^{-14}	3×10^{-2}
$O(^3P)$	8×10^4	6×10^{-14}	2×10^{-3}
RO_2	3×10^9	$<1 \times 10^{-18}$	$<1 \times 10^{-3}$
HO_2	1×10^9	$<1 \times 10^{-18}$	$<4 \times 10^{-4}$
O_3	2.5×10^{12}	$<8 \times 10^{-24}$	$<7 \times 10^{-6}$

Finlayson-Pitts and Pitts (1986) (as cited and adapted by Harrison 2001)
[a]Concentrations typical of a moderately polluted atmosphere

4.1.3 Oxidation: Heterogeneous Gas-Solid Phase

The surface area of mineral dust particles can facilitate catalysis of SO_2 oxidation (Usher et al. 2002). Gaseous SO_2 irreversibly adsorbs to particles as sulfite and bisulfite and may then be oxidized by O_3 (Usher et al. 2002), H_2O_2, or other trace gases (Huang et al. 2016). These reactions are dependent upon pH, catalyst concentrations, and oxidant concentrations (Beilke and Gravenhorst 1978). Further, high relative humidity may increase the importance of these reactions (Huang et al. 2016). Particles that have been oxidized to sulfate are hygroscopic and form an aqueous layer that may then react further with SO_2 (Usher et al. 2002). However, this mechanism may be of more importance in urban environments where heavy metal concentrations are higher (Beilke and Gravenhorst 1978).

4.1.4 Volatilization and Inhalation Toxicity

The high vapor pressure of SO_2 suggests that it will tend to partition in the air (Table 1) and its formulation as a gas suggests potential off-site movement (US EPA 2013). SO_2 and other sulfur oxide gases (SO_x) are important atmospheric contaminants because of their direct health impacts and indirect impacts due to the formation of sulfate particles, which may increase visibility impairment and contribute to particulate matter or acid rain (US EPA 2016a). Humans and terrestrial nontarget organisms including invertebrates, mammals, birds, and plants may be exposed to SO_2 by post-fumigation releases of SO_2 into the environment (US EPA 2013). The US EPA requested that the registrants of SO_2 products submit special studies on the impacts of SO_2 on terrestrial organisms by June of 2016 (US EPA 2014a), including a honeybee inhalation study, an avian inhalation study, and a terrestrial plant study of vegetation vigor (US EPA 2013). A common measure of chemical potency is the LC_{50}, which is the concentration of a chemical at which exposure for a specific duration of time results in 50% mortality of experimental laboratory animals (ATSDR 1998). Table 5 includes LC_{50} values for inhalation exposures to SO_2 in various mammals.

Table 5 Toxicity tests of mammalian inhalation exposure to sulfur dioxide (US EPA 2007)

Species	LC_{50} (ppm)	Duration
Mouse	3,000	30 min
Mouse	1,000	4 h
	150	847 h (~35 days)
Rat	2,520	1 h
	2,168	5 h
Guinea pig	1,039	24 h
Guinea pig	1,000	20 h
	130	154 h (~6 days)

4.1.5 Wet and Dry Deposition

Wet deposition is the combination of removal processes through which hydrometeors (rain, snow, fog, etc.) scavenge materials from the air, which is facilitated by falling precipitation ("washout") or in-cloud ("rainout") processes (Alberta Environment 2003; Seinfeld and Pandis 2016). SO_2 is estimated to have an atmospheric lifetime of approximately 7 days with respect to wet deposition; however, the highly variable nature of precipitation greatly influences rates of wet deposition (US EPA 2008a). The size distribution of cloud droplets, rain droplets, and aerosols also influences these rates (Seinfeld and Pandis 2016).

Dry deposition refers to the removal of gases or particles from the atmosphere and transferal to land and sea surfaces without the influence of precipitation (Harrison 2001; Seinfeld and Pandis 2016). Dry deposition is believed to contribute only 15% of the net loss of sulfate from the atmosphere due to wet deposition, although the relative contribution widely varies by region (Faloona 2009; US EPA 2014c). The atmospheric lifetime of SO_2 with respect to dry deposition is between 1 and 7 days (US EPA 2008a).

Deposition velocity is the rate at which substances are expected to deposit from the air to various surfaces (Harrison 2001; Seinfeld and Pandis 2016), as described by Eq. (4). Harrison (2001) listed deposition velocity values typical of different surface types. The estimated deposition velocity of SO_2 over various surfaces ranged from 0.5 to 2.0 cm s^{-1} (ocean < soil < grass < forest). Atmospheric lifetime was shown to be proportional to the boundary layer depth and inversely proportional to deposition velocity (Harrison 2001). However, deposition velocity is a simple representation of a number of more complex processes, including aerodynamic atmospheric transport and molecular transport of gases or Brownian transport of particles across a thin quasilaminar sublayer, followed finally by surface uptake (Seinfeld and Pandis 2016).

Deposition Velocity Equation
The velocity of sulfur dioxide deposition is dependent on the speed of movement of sulfur dioxide particles toward the Earth's surface and particle concentration (Harrison 2001; Seinfeld and Pandis 2016).

$$V_g = \frac{F}{C} \qquad (4)$$

V_g deposition velocity (m s^{-1}), F flux to surface (μg m^{-2} s^{-1}), C atmospheric concentration (μg m^{-3}).

Acid rain is the combined contribution of wet deposition and dry deposition to the removal of acidic compounds from the atmosphere (Seinfeld and Pandis 2016). In the Eastern United States, acid rain has severely impacted many lakes and forests (Gliessman 2007). The formation of acid rain and acid fog has historically been problematic in Southern California and the San Joaquin Valley, likely due to local atmospheric pollutants (related to transportation or energy extraction, respectively) and meteorological conditions influenced by the surrounding topography (Gliessman 2007; CARB 1983, 1985). Mountain slopes may also experience enhanced acid deposition on slopes where clouds are frequently intercepted, since cloud droplets are generally five to ten times more concentrated than precipitation (Seinfeld and Pandis 2016).

4.2 Environmental Fate and Degradation in Soil

The soil redox state determines the oxidation state of sulfur in soils. In aerobic soils, SO_2 is oxidized to sulfate (SO_4^{2-}). In contrast, SO_2 is reduced to elemental sulfur and sulfides (S^{2-}) in anaerobic soils, producing H_2S gas or thiol-containing organic compounds (Strawn et al. 2015).

Atmospheric SO_2 is oxidized in the atmosphere to H_2SO_4 and contributes to acid rain, which causes widespread ecosystem impacts (Strawn et al. 2015). Plant toxicity is largely due to the mobilization of aluminum, manganese, and iron ions (USDA 2011) by cation exchange in acidified soils (Strawn et al. 2015). Low soil pH and high aluminum ion concentration inhibit the microbial decomposition of plant matter and inhibit the growth of fungi, earthworms, and plants (USDA 2011). This ultimately reduces the availability of essential nutrients (i.e., calcium, magnesium, phosphate, and nitrate) in soils (USDA 2011).

Although acidic soils can be mediated by liming in agricultural settings, natural landscapes such as forests and prairies are vulnerable to soil acidification (Strawn et al. 2015). The capacity of soils to buffer soil acidity by neutralizing acidic rainwater is determined by the type of bedrock and the thickness and composition of the soil (US EPA 2016a). Ecosystems with thin soils and poor neutralizing ability (e.g., mountainous regions) are particularly vulnerable to acidic soils (CARB 2002; US EPA 2016a).

4.2.1 Biodegradation

Once deposited onto soil, SO_2 is either oxidized to sulfates or reduced to sulfide depending upon the availability of oxygen and the microbial community

(Griffith et al. 2015; Eriksen et al. 1998). Therefore, soil factors that affect microbial community composition and activity such as temperature, moisture, and pH also affect these reaction rates (Eriksen et al. 1998). Volatilization of microbially produced reduced sulfur compounds is a minor degradation pathway for sulfur compounds in soils (Havlin et al. 2005). Once volatilized, however, rapid atmospheric oxidation of compounds such as H_2S may produce SO_2 (Moss 1978), which demonstrates processes of sulfur biogeochemical cycling.

4.3   Environmental Fate and Degradation in Water

SO_2 is highly water soluble (Table 1) and upon dissolution results in a slightly acidic solution (Eq. 1) dominated by HSO_3^- within the pH range from approximately 2 to 7 and dominated by SO_3^{2-} at higher pH (Seinfeld and Pandis 2016). Solubility of SO_2 is increased at higher pH levels (Harrison 2001).

Sulfate ions are readily transferred from soils to surface waters and are very mobile in the environment (Mason 2001). Farms may be contaminated by contributions of acid rain runoff to irrigation water, lowering pH and potentially impacting crops (USDA 2011). The National Surface Water Survey has shown that many lakes and streams suffer from chronic acidity, which has negative impacts on ecosystems (USDA 2011). Episodic acidification can occur when soils lack buffering capacity to prevent acidification caused by snowmelt or heavy precipitation (US EPA 2016a). These exposures to high acidity may cause physiological stress, injury, or mortality in various organisms (US EPA 2016a).

4.3.1   Aquatic Organisms

Acidification of surface waters due to acid rain has been investigated over several decades, with observations of fish declines reported as early as the 1920s in Scandinavian lakes (Mason 2001). Freshwater acidification causes various negative impacts to fish, such as individual mortality, decreased population size, population extirpation, and decreased biodiversity (USDA 2011). The reduced biodiversity of aquatic ecosystems resulting from acidification may extend up higher trophic levels, including birds and mammals (Mason 2001). Acid deposition may be particularly harmful to lakes banked by poorly buffered soils over granitic bedrock (Seinfeld and Pandis 2016). Acidification may be chronic; however, acidification may also be episodic, for example, if snowmelt or downpours temporarily decrease pH beyond soil buffering capacity (CARB 2002; US EPA 2016a).

Fish and invertebrate mortality is caused by interference in the normal ionic equilibria of essential ions including sodium, chloride, potassium, and calcium, as well as increased uptake of mobilized aluminum ions (Mason 2001). Acid rain runoff can leach aluminum from soil clay particles, which may then flow into surface waters (US EPA 2016a). Sulfate-mediated acidification of aquatic systems can also mobilize other toxic metals such as lead and cadmium (Griffith et al. 2015). Acidification is particularly harmful for organisms in developmental stages and may prevent normal growth and population recruitment (Mason 2001). Sulfate deposition can also increase mercury methylation rates in wetlands and aquatic ecosystems and potentially increase concentrations of mercury in fish that may be consumed by adults, pregnant women, or children (US EPA 2010).

5 Impacts to Vegetation and Crops

SO_2 can damage plants and crops by causing foliar injury and decreasing growth, yield, and plant diversity within a given community (US EPA 2014b). Impacts on plant growth may occur directly from SO_2 or indirectly through changes to soil systems (USDA 2011). Plant leaves can absorb SO_2 from the atmosphere and have been found to contain elevated levels of sulfate near sources of atmospheric SO_2 emissions (Al-Jahdali and Bisher 2008).

Relatively low concentrations of SO_2 have resulted in foliar injury (0.5 ppm) and severe stress (1–2 ppm) to plants (Havlin et al. 2005; US EPA 2007). In polluted areas, the absence of lichens has been used as a bioindicator of SO_2 pollution (US EPA 2014b). Greenhouse experiments simulating low pH exposure from California acid fogs have indicated that crop injury and increased disease suscepti-bility may occur (Musselman et al. 1988).

Wet deposition of SO_2 as acid rain may affect seedling germination and may also damage the tissues and waxy coatings of the leaves and needles of plants (Gliessman 2007). Trees and whole forests have been impacted by acid rain, which is a major environmental problem in the Northern Hemisphere (US EPA 2013). However, enforcement of the Federal Clean Air Act has resulted in substantial decreases in measured ambient SO_2 levels (US EPA 2016b). For example, there was a 69% decrease in average national SO_2 levels and a 48% decrease in average regional (California and Nevada) SO_2 levels from 2000 to 2015 (US EPA 2016b).

6 Impacts to Man-Made Materials

Deposition of SO_2 can accelerate corrosion of metals, concrete, limestone, and other materials due to the formation of H_2SO_4 (US EPA 2016a). Acid rain impacts structures including buildings, statues, and monuments (US EPA 2010). Such effects are considered public welfare impacts (US EPA 2008a).

7 Persistence

Based upon reaction with the hydroxyl radical, the atmospheric lifetime of SO_2 has been estimated as approximately 7 days, with similar estimates (1–7 days) for dry deposition (US EPA 2008a). Precipitation may decrease the atmospheric lifetime of SO_2 due to oxidation and deposition processes, resulting in an atmospheric lifetime on the order of days (Griffith et al. 2015; US EPA 2008a). Convection-driven vertical transport of SO_2 to the upper atmosphere can result in longer atmospheric lifetimes and greater long-distance transport of atmospheric SO_2 from source areas (US EPA 2008a).

8 Environmental Monitoring

Ambient air monitoring of SO_2 as a criteria pollutant has been conducted nationwide since 1979 (US EPA 2010). Monitoring is performed via an SO_2 monitoring network including State and Local Air Monitoring Stations (SLAMS) and National Air Monitoring Stations (NAMS) at approximately 488 sites nationwide (US EPA 2010). These ambient air monitoring programs are operated primarily by state and local agencies to compare measured air concentrations to the NAAQS and to make air pollution data available to the public, among other research objectives (US EPA 2010).

US EPA (2008a) SO_2 monitoring network data from 2003 to 2005 showed a gradient of increasing SO_2 concentrations from the West to the East Coast of the United States. In the 12 metropolitan areas with at least 4 SO_2 air monitoring stations, reported mean annual concentrations ranged from approximately 1 ppb in Riverside and San Francisco, CA, to 12 ppb in Pittsburgh, PA, and 14 ppb in Steubenville, OH. During this period, the annual average concentration was 4 ppb, with a maximum value of >700 ppb and 1-h maximum average concentrations of 13 ppb. Estimated background concentrations of SO_2 are relatively small (<10–30 ppt) and are estimated to contribute <1% of total ambient SO_2 concentrations in the United States. However, on the Northwest Coast, areas of high volcanic activity may contribute to up to 70–80% of ambient SO_2 concentrations, although concentrations are typically below 2 ppb in these areas (US EPA 2008a).

According to CARB (2011a), California has been in attainment with SO_2 standards since the late 1980s, with concentrations decreasing from as high as 230 ppb in the 1970s to less than 50 ppb in the 1990s. In 2009, 1-h SO_2 air concentrations (ranging from 3 to 35 ppb) were reported at approximately one-tenth of the concentrations measured in the 1970s. CARB estimates that SO_2 comprises 97% of the SO_x emissions detected; therefore, SO_x emissions are presented as emissions of SO_2 (CARB 2011a). Appendix, Table 10 summarizes the average annual 99th percentile of the 1-h daily maximum SO_2 concentrations detected from 2007 to 2009 (CARB 2011b). Since 1990, emissions of SO_2 have decreased by 45% (CARB 2011a). The highest concentrations of SO_2 were detected in the South Central Coast, San Francisco Bay Area, and South Coast Air Basins (CARB 2011a).

Additionally, various Eulerian and Lagrangian atmospheric chemical transport models are used to better understand atmospheric processes, interpret monitoring results, or make regulatory decisions, as each approach to characterize ambient SO_2 concentrations is restricted by inherent uncertainties and limitations (Seinfeld and Pandis 2016; US EPA 2008c). Lagrangian models calculate air concentrations within an air parcel as it moves in space and time, whereas Eulerian models calculate air concentrations of a grid of air parcels that remain fixed in space over time (Seinfeld and Pandis 2016).

9 Conclusion

SO_2 is a moderately persistent, highly water soluble atmospheric pollutant that will tend to partition to the atmosphere where it may be transported, deposited, or transformed. The main degradation route of SO_2 is atmospheric oxidation, and sulfur oxides may undergo long-distance transport until removed from the atmosphere by wet or dry deposition. SO_2 used as a pesticide will enter the sulfur cycle, in which abiotic and biotic reactions cycle sulfur-containing compounds between the environment and living organisms. SO_2 is used as a fungicide for cold storage of post-harvest grape products and in wineries to prevent mold growth and is also used as an antimicrobial to sterilize wine barrels and other equipment. In recent years, there has been a slight increasing trend in both the acreage of grapes and the amount of SO_2 used as a pesticide in California. Although agricultural contributions of SO_2 emissions are estimated to be minimal compared to anthropogenic emissions from fossil fuels, SO_2 emissions from pesticide uses may contribute to the negative environmental and public welfare impacts of acid rain resulting from the oxidation of atmospheric SO_2 to sulfur oxides. The negative impacts of acid rain include toxicity to aquatic organisms and fish, toxicity to terrestrial vegetation, and increased corrosion of manmade materials.

10 Summary

In California, uses of SO_2 as a pesticide from 2010 to 2015 were primarily for fumigations (96%), including treatments of post-harvest grape products and winery equipment sterilizations. The highest reported total monthly use of SO_2 in California was observed during the months of July–November. The total annual use of SO_2 as a pesticide increased from 2010 to 2015, which may correspond to an increase in the reported acreage of grapes in California during the same time period. Although the primary sources of atmospheric SO_2 are anthropogenic emissions from the combustion of fossil fuels, SO_2 emissions from pesticide uses of SO_2 have the potential to contribute to the impacts of SO_2 pollution. Atmospheric SO_2 participates in the sulfur biogeochemical cycle, which involves reactions between sulfur-containing compounds that cycle between abiotic and biotic components of ecosystems. The oxidation of atmospheric SO_2 to sulfur oxides may contribute to the environmental, public welfare, and other impacts of SO_2 pollution.

Acknowledgments The author would like to thank the scientists at the California Department of Pesticide Regulation, with special acknowledgment to Minh Pham, Edgar Vidrio, Randy Segawa, Pam Wofford, and Madeline Brattesani.

Disclaimer The mention of commercial products, their source, or their use in connection with material reported here is not to be construed as either an actual or implied endorsement of such products.

Conflict of Interest The author has no conflicts of interest to declare; however, it should be noted that DPR is the author's employer and the source of the PUR records analyzed within this report. The findings and conclusions of this report are those of the author and do not necessarily represent the views of DPR.

Appendix

Table 6 Sulfur dioxide product formulations and uses

Product (DPR registration no.)	Formulation	% SO_2	Other ingredients (%)	Uses	Reentry level[a] (ppm)
Airgas sulfur dioxide (89867-2-AA)	Pressurized liquid, sprays, foggers	99.9	Inert ingredients (0.1%)	Wine barrel and cork sanitizer	2
The fruit doctor (11195-1-AA)	Pressurized liquid	100	–	Wine barrel and cork sanitizer, postharvest grape fungicide (cold-storage rooms/fumigation chambers or refrigerated trucks/containers/railcars)	2

DPR (2017b), US EPA (2016d)
[a]Air concentration below which no respiratory protection device required

Table 7 Sulfur dioxide product application rates and requirements for grape fumigation

Treatment type	Minimum treatment interval (days)	Minimum hold time prior to shipment (if gassed >3 times) (h)	Initial fumigation concentration (%)	Maintenance fumigation concentration (%)	Grape variety	Maximum number of treatments
Pre-ship-ment storage	7–10	12	0.75–1	0.25–0.5	Seeded	20
					Seedless (except Thompson seedless)	15
					Thompson seedless	12
Extended storage	2–3	12	0.75–1	0.02–0.04	Seeded	NA
					Seedless	NA

"The Fruit Doctor" product label (Snowden Enterprises 2016)

Table 8 Total cumulative pesticide use of sulfur dioxide (pounds of active ingredient[a]) reported in California from 2010 to 2015, summarized by county

County	2010	2011	2012	2013	2014	2015	Annual average	Total	% of total
Kern	66,420	73,330	54,200	92,290	69,150	64,590	70,000	419,980	31.9
Tulare	61,250	49,290	48,320	68,380	71,650	96,590	65,910	395,480	30.0
Fresno	17,210	25,380	17,200	32,610	31,870	27,680	25,330	151,950	11.5
Napa	14,250	24,780	25,850	22,340	25,430	23,460	22,690	136,110	10.3
Sonoma	14,660	17,840	17,020	15,390	13,520	13,890	15,390	92,320	7.0
Mendocino	2,570	14,020	3,240	1,340	1,640	2,870	4,280	25,680	1.9
Madera	3,720	2,340	2,400	3,130	3,020	4,850	3,240	19,460	1.5
Monterey	1,300	2,930	3,370	2,640	3,490	4,710	3,070	18,440	1.4
Stanislaus	1,350	3,820	1,100	3,000	2,710	3,190	2,530	15,170	1.2
San Luis Obispo	2,120	4,570	2,010	1,730	1,880	2,380	2,450	14,690	1.1
Solano	3,180	1,510	350	1,360	1,170	0	1,260	7,570	0.6
Santa Barbara	760	790	1,270	940	640	950	890	5,350	0.4
All other counties	2,550	5,420	1,700	1,960	1,800	2,850	2,710	16,280	1
Statewide total	191,340	226,020	178,030	247,110	227,970	248,010	219,750	1,318,480	100

DPR (2017d)

[a]Rounded to the nearest 10 pounds

Table 9 Total cumulative pesticide use of sulfur dioxide (pounds of active ingredient[a]) reported in California from 2010 to 2015, summarized by crop or site

Crop or site use	2010	2011	2012	2013	2014	2015	Annual average	Total	% of total
Grapes	89,750	119,060	89,570	115,720	111,300	173,650	116,510	699,050	53.02
Fumigation, other	42,810	74,980	57,460	56,570	49,630	58,390	56,640	339,840	25.78
Commodity fumigation	43,890	2,050	23,170	39,940	54,950	6,960	28,490	170,960	12.97
No crop name indicated	3,490	22,350	0	19,800	350	8,850	9,140	54,840	4.16
Structural pest control	10,720	5,040	1,940	14,370	11,620	10	7,290	43,700	3.31
Landscape maintenance	50	30	4,830	560	30	40	920	5,540	0.42
Public health pest control	<10	2,000	0	0	0	0	330	2,000	0.15
Grapes, wine	310	100	820	150	80	40	250	1,500	0.11
Regulatory pest control	290	230	<10	<10	10	30	90	560	0.04
Research commodity	0	50	200	0	0	0	40	250	0.02
Vertebrate pest control	0	120	0	0	0	0	20	120	0.01
Commercial, institutional, or industrial areas	30	10	30	<10	0	20	20	90	0.01
Rights-of-way	0	<10	<10	0	0	<10	<10	<10	<0.001
Rangeland (all or unspecified)	0	0	0	0	<10	0	<10	<10	<0.001
Statewide total	191,340	226,020	178,030	247,110	227,970	248,010	219,750	1,318,480	100

DPR (2017d)

[a]Rounded to the nearest 10 pounds

Table 10 Sulfur dioxide 1-h federal design values (continued on next page)[a, b]

Air basin	County	Site name	SO2 1-h federal design value (ppb)
Mojave Desert	San Bernardino	Trona-Athol and Telegraph	10
		Victorville-14306 Park Avenue	6
North Coast	Humboldt	Eureka-Jacobs	5
North Central Coast	Santa Cruz	Davenport	11[c]
Sacramento Valley	Sacramento	North highlands-Blackfoot Way	4[c]
		Sacramento-del Paso Manor	4[c]
Salton Sea	Imperial	Calexico-Ethel Street	10
San Diego	San Diego	Chula Vista	7
		Otay Mesa-Paseo International	22[c]
		San Diego-1110 Beardsley Street	17
San Joaquin Valley	Fresno	Fresno-1st Street	9[c]
San Francisco Bay Area	Alameda	Berkeley-6th Street	12[c]
		Oakland-West	13[c]
	Contra Costa	Bethel Island Road	8
		Concord-2975 Treat Blvd.	14
		Crockett-Kendall Avenue	25
		Martinez-Jones Street	18
		Pittsburg-10th Street	20[c]
		Richmond-7th Street	18
		San Pablo-Rumrill Blvd.	14[c]
	San Francisco	San Francisco-Arkansas Street	15[c]
	Santa Clara	San Jose-Jackson Street	5[c]
	Solano	Benicia-East 2nd Street	26[c]
		Vallejo-304 Tuolumne Street	8
South Coast	Los Angeles	Burbank-W. Palm Avenue	8[c]
		Los Angeles-North Main Street	7
		Los Angeles-Westchester Parkway	15[c]
		North Long Beach	20
	Orange	Costa Mesa-Mesa Verde Drive	7
	Riverside	Riverside-Rubidoux	7
	San Bernardino	Fontana-Arrow Highway	6
South Central Coast	San Luis Obispo	Nipomo-Guadalupe Road	35

(continued)

Table 10 (continued)

Air basin	County	Site name	SO2 1-h federal design value (ppb)
	Santa Barbara	El Capitan Beach	4
		Exxon Site 10-UCSB West Campus	4
		Goleta-Fairview	3[c]
		Las Flores Canyon #1	6
		Lompoc-HSandP	2
		Lompoc-S H Street	4
		Vandenberg Air Force Base-STS Power	3

[a]CARB (2011b)

[b]Data were extracted on November 23, 2010 from AQMIS Merged. The 2009 SO_2 1-h federal design values were calculated based on the 3-year average of the annual 99th percentile of the 1-h daily maximum concentrations (2007, 2008, 2009). The federal 1-h SO_2 standard is 75 ppb and was effective August 23, 2010. All SO_2 1-h federal design values in California are below the standard of 75 ppb

[c]Sites do not meet the US EPA/s completeness criteria. No monitoring data are available for the following air basins: Great Basin Valleys, Lake County, Lake Tahoe, Mountain Counties, Northeast Plateau

References

Alberta Environment (2003) Sulphur dioxide: environmental effects, fate and behaviour. Alberta Environment Science and Standards Branch. http://aep.alberta.ca/air/legislation/ambient-air-quality-objectives/documents/SulphurDioxideEffectsFateBehaviour-2003.pdf.   Accessed 3 May 2017

Al-Jahdali MO, Bisher ASB (2008) Sulfur dioxide (SO_2) accumulation in soil and plant's leaves around an oil refinery: a case study from Saudi Arabia. Am J Environ Sci 4(1):84–88. https://doi.org/10.3844/ajessp.2008.84.88

Ashar NG (2016) Chemical and physical properties of sulphur dioxide and sulphur trioxide. In: Advances in sulphonation techniques, Springer briefs in applied sciences and technology. Springer, Berlin. https://doi.org/10.1007/978-3-319-22641-5_2

ATSDR (1998) Toxicological profile for sulfur dioxide. US Department of Health and Human Services, Public Health Service, Agency for Toxic Substances and Disease Registry. https://www.atsdr.cdc.gov/toxprofiles/tp116.pdf. Accessed 3 May 2017

Beilke S, Gravenhorst G (1978) Heterogeneous SO_2-oxidation in the droplet phase. Atmos Environ 12(1–3):231–239. https://doi.org/10.1016/0004-6981(78)90203-2

Blanco G, Gerlagh R, Suh S, Barrett J, de Coninck HC, Diaz Morejon CF, Mathur R, Nakicenovic N, Ofosu Ahenkora A, Pan J, Pathak H, Rice J, Richels R, Smith SJ, Stern DI, Toth FL, Zhou P (2014) Drivers, trends and mitigation. In: Edenhofer O, Pichs-Madruga R, Sokona Y, Farahani E, Kadner S, Seyboth K, Adler A, Baum I, Brunner S, Eickemeier P, Kriemann B, Savolainen J, Schlömer S, von Stechow S, Zwickel T, Minx JC (eds) Climate change 2014: mitigation of climate change. Contribution of Working Group III to the fifth assessment report of the Intergovernmental Panel on Climate Change, pp 351–411. https://www.ipcc.ch/pdf/assessment-report/ar5/wg3/ipcc_wg3_ar5_chapter5.pdf. Accessed 29 May 2017

Bombrun H, Sumner D (2003) What determines the price of wine? The value of grape characteristics and wine quality assessments. University of California Agriculture and Natural Resources, Agricultural Issues Center Brief. http://aic.ucdavis.edu/pub/briefs/brief18.pdf. Accessed 2 Aug 2017

Boubel RW, Fox DL, Turner DB, Stern AC (1994) Fundamentals of air pollution, 3rd edn. Academic Press, San Diego

Bunce N (1994) Environmental Chemistry, 2nd edn. Wuerz Publishing Ltd., Winnipeg, 376 p

CARB (1983) Characterization of reactants, reaction mechanisms and reaction products leading to extreme acid rain and acid aerosol conditions in southern California. California Environmental Protection Agency, California Air Resources Board. https://www.arb.ca.gov/research/apr/past/a0-141-32.pdf. Accessed 9 May 2017

CARB (1985) Characterization of reactants, reaction mechanisms and reaction products in atmospheric water droplets: fog, cloud, dew, and rainwater chemistry. California Environmental Protection Agency, California Air Resources Board. https://www.arb.ca.gov/research/apr/past/a2-048-32a.pdf. Accessed 9 May 2017

CARB (2002) Episodic acidification of lakes in the Sierra Nevada. California Environmental Protection Agency, California Air Resources Board. https://www.arb.ca.gov/research/apr/past/a132-048.pdf. Accessed 4 Apr 2017

CARB (2011a) Recommended area designations for the 2010 federal sulfur dioxide (SO2) standard: staff report. California Environmental Protection Agency, California Air Resources Board. https://www.arb.ca.gov/desig/so2e1.pdf. Accessed 15 Nov 2017

CARB (2011b) SO_2 1-h design values for all sites in California. California Environmental Protection Agency, California Air Resources Board. https://www.arb.ca.gov/desig/so2a2.pdf. Accessed 15 Nov 2017

CDFA (2014) California grape acreage report, 2013 summary. California Department of Food and Agriculture. https://www.nass.usda.gov/Statistics_by_State/California/Publications/Fruits_and_Nuts/2014/201403grpac.pdf. Accessed 24 Apr 2017

CDFA (2015) California grape acreage report, 2014 summary. California Department of Food and Agriculture. https://www.nass.usda.gov/Statistics_by_State/California/Publications/Fruits_and_Nuts/2015/201503grpac.pdf. Accessed 24 Apr 2017

CDFA (2016) California grape acreage report, 2015 summary. California Department of Food and Agriculture. https://www.nass.usda.gov/Statistics_by_State/California/Publications/Fruits_and_Nuts/2016/201604grpac.pdf. Accessed 24 Apr 2017

Cox RA (1973) Particle formation from homogeneous reactions of sulphur dioxide and nitrogen dioxide. Tellus 26(1–2):235–240. http://journals.co-action.net/index.php/tellusa/article/viewFile/9782/11401

Craig K (2017) Advanced processing of pesticide use reports for data analysis conducted by the Environmental Monitoring Branch's Air Program. Memorandum dated October 12, 2017 to Edgar Vidrio. California Environmental Protection Agency, Department of Pesticide Regulation. http://www.cdpr.ca.gov/docs/emon/pubs/ehapreps/analysis_memos/pur_memorandum_final.pdf. Accessed 17 Nov 2017

Dämmgen U, Walker K, Grünhage L, Jäger H (1998) The atmospheric sulphur cycle. In: Schnug E (ed) Sulphur in agroecosystems. Kluwer Academic, Dordrecht, pp 39–73

DPR (2009) Sulfur dioxide use in wineries. California Environmental Protection Agency, Department of Pesticide Regulation. http://www.cdpr.ca.gov/docs/county/cacltrs/penfltrs/penf2009/2009atch/attach1201.pdf. Accessed 3 May 2017

DPR (2011a) Registration for special local need for distribution and use only within California. California Environmental Protection Agency, Department of Pesticide Regulation. http://www.cdpr.ca.gov/docs/label/pdf/sln/244770.pdf. Accessed 30 May 2017

DPR (2011b) Section 24(c): special local need registrations. California Environmental Protection Agency, Department of Pesticide Regulation. http://cdpr.ca.gov/docs/registration/guides/section24c.pdf. Accessed 30 May 2017

DPR (2016a) What you need to know about winery use of sulfur dioxide. California Environmental Protection Agency, Department of Pesticide Regulation. http://www.cdpr.ca.gov/docs/dept/factshts/so2.pdf. Accessed 3 May 2017

DPR (2016b) Summary of pesticide use report data 2015: indexed by commodity. California Environmental Protection Agency, Department of Pesticide Regulation. http://www.cdpr.ca.gov/docs/pur/pur15rep/comrpt15.pdf. Accessed 10 May 2017

DPR (2017a) Pesticide registration. In: A guide to pesticide regulation in California. California Environmental Protection Agency, Department of Pesticide Regulation. http://www.cdpr.ca.gov/docs/pressrls/dprguide/chapter3.pdf. Accessed 3 May 2017

DPR (2017b) Product/label database. California Environmental Protection Agency, Department of Pesticide Regulation. Database queried by DPR staff on March 10, 2016. http://www.cdpr.ca.gov/docs/label/labelque.htm

DPR (2017c) Pesticide Use Reporting. In: a guide to pesticide regulation in California. California Environmental Protection Agency, Department of Pesticide Regulation. http://www.cdpr.ca.gov/docs/pressrls/dprguide/chapter9.pdf. Accessed 3 May 2017

DPR (2017d) Pesticide Use Reporting (PUR). California Environmental Protection Agency, California Department of Pesticide Regulation. PUR Database queried by DPR Staff on August 2, 2017. http://www.cdpr.ca.gov/docs/pur/purmain.htm. Accessed 2 Aug 2017

Eggleton AEJ, Cox RA (1978) Homogeneous oxidation of sulphur compounds in the atmosphere. Atmos Environ 12(1–3):227–230. https://doi.org/10.1016/0004-6981(78)90202-0

Encyclopedia Britannica (2008) The sulfur cycle. Encyclopedia Britannica Online. https://www.britannica.com/media/full/572740/111671. Accessed 4 May 2017

Eriksen J, Murphy MD, Schnug E (1998) The soil sulphur cycle. In: Schnug E (ed) Sulphur in agroecosystems. Kluwer Academic, Dordrecht, pp 39–73

Faloona I (2009) Sulfur processing in the marine atmospheric boundary layer: a review and critical assessment of modeling uncertainties. Atmos Environ 43(18):2841–2854. https://doi.org/10.1016/j.atmosenv.2009.02.043

Finlayson-Pitts BJ, Pitts JN Jr (1986) Acid deposition. In: Atmospheric chemistry: fundamentals and techniques. Wiley, New York, pp 645–693

Forster P, Ramaswamy V, Artaxo P, Berntsen T, Betts R, Fahey DW, Haywood J, Lean J, Lowe DC, Myhre G, Nganga J, Prinn R, Raga G, Schulz M, Van Dorland R (2007) Changes in atmospheric constituents and in radiative forcing. In: Solomon S, Qin D, Manning M, Chen Z, Marquis M, Averyt KB, Tignor M, Miller HM (eds) Climate change 2007: the physical science basis. Contribution of Working Group I to the fourth assessment report of the Intergovernmental Panel on Climate Change, pp 129–234. https://www.ipcc.ch/publications_and_data/ar4/wg1/en/ch2s2-4-4-1.html. Accessed 29 May 2017

Gammon DW, Moore TB, O'Malley MA (2010) Toxicology of sulfur dioxide. In: Krieger R, Doull J, Hodgson E, Mmaibach H, Reiter L, Ritter L, Ross J, Slikker W Jr, von Hemmen J (eds) Hayes' handbook of pesticide toxicology, 3rd edn. Elsevier, London, pp 1889–1900

Gliessman S (2007) Humidity and rainfall. In: Engles E (ed) Agroecology: the ecology of sustainable food systems. CRC Press, Taylor & Francis, Boca Raton, pp 67–79

Griffith CM, Woodrow JE, Seiber JN (2015) Environmental behavior and analysis of agricultural sulfur. Pest Manag Sci 71(11):1486–1496. https://doi.org/10.1002/ps.4067

Harris E, Sinha B, van Pinxteren D, Schneider J, Poulain L, Collett J, D'Anna B, Fahlbusch B, Foley S, Fomba KW, George C, Gnauk T, Henning S, Lee T, Mertes S, Roth A, Stratmann F, Borrmann S, Hoppe P, Herrmann H (2014) In-cloud sulfate addition to single particles resolved with sulfur isotope analysis during HCCT-2010. Atmos Chem Phys 14(8):4219–4235. https://doi.org/10.5194/acp-14-4219-2014

Harrison RM (2001) Chemistry and climate change in the troposphere: atmospheric acids. In: Harrison RM (ed) Pollution: causes, effects, and control. Royal Society of Chemistry, Cambridge, pp 92–95. https://doi.org/10.1039/9781847551719-FP001

Havlin JL, Beaton JD, Tisdale SL, Nelson WL (2005) Sulfur, calcium, and magnesium. In: Yarnell D (ed) Soil fertility and fertilizers: an introduction to nutrient management. Pearson Prentice Hall, Upper Saddle River, pp 219–243

Hoyle CR, Fuchs C, Järvinen E, Saathoff H, Dias A, El Haddad I, Gysel M, Coburn SC, Tröstl J, Bernhammer AK, Bianchi F, Breitenlechner M, Corbin JC, Craven J, Donahue NM, Duplissy J, Ehrhart S, Frege C, Gordon H, Höppel N, Heinritzi M, Kristensen TB, Molteni U, Nichman L, Pinterich T, Prévôt ASH, Simon M, Slowik JG, Steiner G, Tomé A, Vogel AL, Volkamer R, Wagner AC, Wagner R, Wexler AS, Williamson C, Winkler PM, Yan C, Amorim A, Dommen J, Curtius J, Gallagher MW, Flagan RC, Hansel A, Kirkby J, Kulmala M, Möhler O, Stratmann F, Worsnop DR, Baltensperger U (2016) Aqueous phase oxidation of sulphur dioxide by ozone in cloud droplets. Atmos Chem Phys 16(3):1693–1712. https://doi.org/10.5194/acp-16-1693-2016

Huang L, Zhao Y, Li H, Chen Z (2016) Hydrogen peroxide maintains the heterogeneous reaction of sulfur dioxide on mineral dust proxy particles. Atmos Environ 141:552–559. https://doi.org/10.1016/j.atmosenv.2016.07.035

Langner J, Rodhe H (1991) A global three-dimensional model of the tropospheric sulfur cycle. J Atmos Chem 13(3):225–263. https://doi.org/10.1007/BF00058134

Liang J, Jacobson MZ (1999) A study of sulfur dioxide oxidation pathways over a range of liquid water contents, pH values, and temperatures. J Geophys Res 104(D11):13749–13769. https://doi.org/10.1029/1999JD900097

Mason CF (2001) Water pollution biology: acidification. In: Harrison RM (ed) Pollution: causes, effects, and control. Royal Society of Chemistry, Cambridge, pp 92–95. https://doi.org/10.1039/9781847551719-FP001

Moss MR (1978) Sources of sulfur in the environment: the global sulfur cycle. In: Metcalf RL, Pitts JN, Stumm W (eds) Sulfur in the environment; part I: the atmospheric cycle, Environmental science and technology. Wiley, New York, pp 23–50

Musselman RC, McCool P, Sterret JL (1988) Acid fog injures California crops. Calif Agric 42(4): 6–7. https://ucanr.edu/repositoryfiles/ca4204p6-68791.pdf. Accessed 29 May 2017

Newland MJ, Rickard AR, Mohammed SA, Vereecken L, Muñoz A, Ródenas M, Bloss WJ (2015) Kinetics of stabilized criegee intermediates derived from alkene ozonolysis: reactions with SO_2, H_2O and decomposition under boundary layer conditions. Phys Chem Chem Phys 12(6):4076–4088. https://doi.org/10.1039/C4CP04186K

Nriagu J (1978) Production and uses of sulfur. In: Metcalf RL, Pitts JN, Stumm W (eds) Sulfur in the environment; part I: the atmospheric cycle, Environmental science and technology. Wiley, New York, pp 1–21

Ruth JH (1986) Odor thresholds and irritation levels of several chemical substances: a review. Am Ind Hyg Assoc J 47:A142–A151. https://www.ncbi.nlm.nih.gov/pubmed/3706135

Sander R (2015) Compilation of Henry's law constants (version 4.0) for water as solvent. Atmos Chem Phys 15(8):4399–4981. https://doi.org/10.5194/acp-15-4399-2015

Sander SP, Abbatt J, Barker JR, Burkholder JB, Friedl RR, Golden DM, Huie RE, Kolb CE, Kurylo MJ, Moortgat GK, Orkin VL, Wine PH (2011) Heterogeneous chemistry. In: Chemical kinetics and photochemical data for use in atmospheric studies: evaluation no. 17. JPL Publication 10-6. National Aeronautics and Space Administration (NASA), Jet Propulsion Laboratory, California Institute of Technology, pp 535–616. http://jpldataeval.jpl.nasa.gov/pdf/JPL%2010-6%20Final%2015June2011.pdf

Sarwar G, Simon H, Fahey K, Mathur R, Goliff W, Stockwell WR (2014) Impact of sulfur dioxide oxidation by stabilized criegee intermediate on sulfate. Atmos Environ 85:204–214. https://doi.org/10.1016/j.atmosenv.2013.12.013

Seinfeld JH, Pandis SN (2016) Atmospheric chemistry and physics: from air pollution to climate change, 3rd edn. Wiley, Hoboken. doi: https://doi.org/10.1021/ja985605y

Smith SJ, Pitcher H, Wigley TML (2001) Global and regional anthropogenic sulfur dioxide emissions. Glob Planet Chang 29(1–2):99–119. https://doi.org/10.1016/S0921-8181(00)00057-6

Smith SJ, Aardenne JV, Klimont Z, Andres RJ, Volke A, Delgado Arias S (2011) Anthropogenic sulfur dioxide emissions: 1850–2005. Atmos Chem Phys 11(3):1101–1116. https://doi.org/10.5194/acp-11-1101-2011

Snowden Enterprises (2016) The fruit doctor: compressed sulfur dioxide (Product label). https://www3.epa.gov/pesticides/chem_search/ppls/011195-00001-20160401.pdf. Accessed 15 May 2017

Strawn DG, Hinrich LB, O'Connor GA (2015) Properties of elements and molecules. In: Soil chemistry, 4th edn. Wiley, Oxford

US EPA (2007) Reregistration eligibility decision – inorganic sulfites. Office of Prevention, Pesticides and Toxic Substances, Office of Pesticide Programs. https://archive.epa.gov/pesticides/reregistration/web/pdf/inorganicsulfites.pdf. Accessed 6 May 2017

US EPA (2008a) Source to dose. In: Integrated science assessment (ISA) for sulfur oxides – health criteria. United States Environmental Protection Agency, Washington, pp 47–112. https://cfpub.epa.gov/ncea/isa/recordisplay.cfm?deid=198843. Accessed 15 May 2017

US EPA (2008b) Introduction. In: Integrated science assessment (ISA) for sulfur oxides – health criteria. United States Environmental Protection Agency, Washington, pp 34–46. https://cfpub.epa.gov/ncea/isa/recordisplay.cfm?deid=198843. Accessed 15 May 2017

US EPA (2008c) Annex B: additional information on the atmospheric chemistry of SOx. In: Integrated science assessment (ISA) for sulfur oxides – health criteria. United States Environmental Protection Agency, Washington, pp 245–262. https://cfpub.epa.gov/ncea/isa/recordisplay.cfm?deid=198843. Accessed 15 May 2017

US EPA (2010) Final Regulatory Impact Analysis (RIA) for the SO2 National Ambient Air Quality Standards (NAAQS). United States Environmental Protection Agency. https://www3.epa.gov/ttnecas1/regdata/RIAs/fso2ria100602full.pdf. Accessed 30 May 2017

US EPA (2012) Estimation programs interface suite™ for Microsoft® Windows, v 4.11. United States Environmental Protection Agency. https://www.epa.gov/tsca-screening-tools/epi-suitetm-estimation-program-interface. Accessed 12 May2017

US EPA (2013) Registration review: preliminary problem formulation for ecological risk, environmental fate, endangered species, and drinking water assessments for sulfur dioxide (PC Code 077601) and sodium metabisulfite (PC Code 111409). United States Environmental Protection Agency, Office of Prevention, Pesticides, and Toxic Substances. https://www.regulations.gov/document?D=EPA-HQ-OPP-2013-0598-0004. Accessed 17 May 2017

US EPA (2014a) Sodium metabisulfite and sulfur dioxide final work plan: registration review case numbers 7019 and 4056. United States Environmental Protection Agency, Office of Pesticide Programs, Washington, pp 1–8. http://www.regulations.gov/#!documentDetail;D=EPA-HQ-OPP-2013-0598-0014. Accessed 30 May 2017

US EPA (2014b) Report on the environment: sulfur dioxide emissions. United States Environmental Protection Agency. https://cfpub.epa.gov/roe/indicator.cfm?i=22. Accessed 17 May 2017

US EPA (2014c) Report on the environment: acid deposition. United States Environmental Protection Agency. https://cfpub.epa.gov/roe/indicator.cfm?i=1. Accessed 17 May 2017

US EPA (2015) 2011 national emissions inventory, version 2 technical support document. United States Environmental Protection Agency. https://www.epa.gov/sites/production/files/2015-10/documents/nei2011v2_tsd_14aug2015.pdf. Accessed 6 May 2017

US EPA (2016a) Effects of acid rain. United States Environmental Protections Agency. https://www.epa.gov/acidrain/effects-acid-rain. Accessed 6 May 2017

US EPA (2016b) Sulfur dioxide trends. United States Environmental Protections Agency. https://www.epa.gov/air-trends/sulfur-dioxide-trends#soreg. Accessed 17 May 2017

US EPA (2016c) Sulfur dioxide. Office of pesticide programs. Reregistration case: inorganic sulfites, case #4056. United States Environmental Protection Agency. https://ofmpub.epa.gov/apex/pesticides/f?p=CHEMICALSEARCH:3:::NO:1,3,31,7,12,25:P3_XCHEMICAL_ID:3969. Accessed 17 May 2017

US EPA (2016d) Pesticide Product Label System (PPLS). United States Environmental Protection Agency. https://iaspub.epa.gov/apex/pesticides/f?p=PPLS:1. Accessed 5 May 2017

USDA (2011) Technical evaluation report: sulfur dioxide-crops. United States Department of
    Agriculture, USDA National Organic Program. https://www.ams.usda.gov/sites/default/files/
    media/Sulfur%20dioxide%20smoke%20bombs%20report%202011.pdf. Accessed 6 May 2017
Usher CR, Al-Hosney H, Carlos-Cuellar S, Grassian VH (2002) A laboratory study of the
    heterogeneous uptake and oxidation of sulfur dioxide on mineral dust particles. J Geophys
    Res 107(D23):ACH16-1–ACH16-9. https://doi.org/10.1029/2002JD002051
Wilson WE (1978) Sulfates in the Atmosphere: A Project Report on Project MISTT. Atmos
    Environ 12(1–3):537–547. https://doi.org/10.1016/0004-6981(78)90235-4

A Nondestructive Method to Identify POP Contamination Sources in Omnivorous Seabirds

Rosanne J. Michielsen, Judy Shamoun-Baranes, John R. Parsons, and Michiel H.S. Kraak

Contents

Abbreviations

DDE	1,1′-(2,2-Dichloro-1,1-ethenediyl)bis(4-chlorobenzene)
DDT	1,1′-(2,2,2-Trichloro-1,1-ethanediyl)bis(4-chlorobenzene)
DecaBDE	Decabromodiphenyl ether
OCP	Organochlorine pesticide
PBDE	Polybrominated diphenyl ether
PCB 118	2,3′,4,4′,5-Pentachlorobiphenyl
PCB 153	2,2′,4,4′,5,5′-Hexachlorobiphenyl
PCB 52	2,2′,5,5′-Tetrachlorobiphenyl

R. J. Michielsen (✉) · J. Shamoun-Baranes · J. R. Parsons · M. H. S. Kraak
Institute for Biodiversity and Ecosystem Dynamics, University of Amsterdam, Amsterdam, The Netherlands
e-mail: J.Z.Shamoun-Baranes@uva.nl; J.R.Parsons@uva.nl; M.H.S.Kraak@uva.nl

© Springer International Publishing AG 2018
P. de Voogt (ed.), *Reviews of Environmental Contamination and Toxicology*
Volume 246, Reviews of Environmental Contamination and Toxicology 246,
DOI 10.1007/398_2018_12

PCB   Polychlorinated biphenyl
POP   Persistent organic pollutant

1 Introduction

Persistent organic pollutants (POPs) are highly bioaccumulative chemicals that are present in almost all environments, despite the ban on the production of most of these substances (Stockholm Convention 2009, 2011, 2013, 2015). Animals that inhabit contaminated environments may contain high concentrations of POPs due to bioaccumulation and biomagnification within a food web, which might lead to an array of adverse effects such as the disruption of their endocrine homeostasis (Gould et al. 1999; MacKay and Fraser 2000; Borgå et al. 2004; Fernie et al. 2005). Recently, Jamieson et al. (2017) reported high concentrations of POPs in arthropods living on the bottom of the Mariana trench, highlighting the global extent of the pollution by these chemicals. Particularly high concentrations of these substances were detected in leachate and dust from landfills (Hansen et al. 1997; Öman and Junestedt 2008; Li et al. 2012, 2014; Melnyk et al. 2015). Hence, animals that inhabit or regularly visit contaminated landfills or other contaminated areas might be exposed to high POP concentrations (Gould et al. 1999; Fernie et al. 2005; Técher et al. 2016). It is therefore alarming that several bird species increasingly forage on landfills and waste treatment areas and have subsequently altered their foraging and even migration behavior. Studies from Canada, Western and Central Europe, and Asia report that landfills are utilized by raptors, gulls (*Larus* sp.) corvids, and white storks (*Ciconia ciconia*) (Baxter and Allan 2006; Elliott et al. 2006; Kruszyk and Ciach 2010; de la Casa-Resino et al. 2014; Patenaude-Monette et al. 2014; Fazari and Mcgrady 2016; Tauler-Ametller et al. 2017). In Western and Central Europe, the concern is increasing that due to the overabundance of anthropogenic food provided by landfills, white storks are short stopping their migration (Blanco 1996; Massemin-Challet et al. 2006; Kruszyk and Ciach 2010; de la Casa-Resino et al. 2014). Similarly, the accessibility of landfills influences the distribution of gull species in Europe (Sol et al. 1995; Arizaga et al. 2014). The harmful effect of POP contamination on the reproductive success of species that live or forage in contaminated areas has been reported for several bird species, like tree swallows (*Tachycineta bicolor*), ring-billed gulls (*Larus delawarensis*), and European starlings (*Sturnus vulgaris*) (Halbrook and Arenal 2003; Gilchrist et al. 2014; Técher et al. 2016). Thus, POPs are very widespread but heterogeneous in their distribution. Hence, in order to take effective measures to mitigate the effects on bird populations, it is important to identify the main sources of POP contamination in bird populations.

Birds have long been suggested to function as suitable monitors of environmental pollutants although drawbacks to using certain species and ethical objections have also been noted (Furness 1993, 1997). Gulls are known for decades to opportunistically utilize anthropogenic resources (Bosch et al. 1994; Belant et al. 1998; Duhem et al. 2003; Christel et al. 2012; Caron-Beaudoin et al. 2013; Scott et al. 2014; van Donk et al. 2017). This behavior has led this species group to be involved in many types of human-wildlife conflicts, like collisions with aircrafts and wind turbines, changing EU policies regarding fishery discards, and landfills and nuisance due to increased urban gull populations (Belant et al. 1993; Dolbeer et al. 1993; Sol et al. 1995; Belant 1997; Garthe and Hüppop 2004; Hüppop et al. 2006; Soldatini et al. 2008; Bernhardt et al. 2010; Bicknell et al. 2013; Abdennadher et al. 2014; Arizaga et al. 2014; Tyson et al. 2015; Sommerfeld et al. 2016). Simultaneously, the successful adaptation to human activities could have adverse effects on gull populations, by enhancing their exposure to harmful substances like POPs (Técher et al. 2016). Many gull populations have been declining during recent years, and this may in part be attributed to the adverse effects of POP contamination (Hario and Rintala 2016; Poprach et al. 2016; Técher et al. 2016). The close connection of gull populations with human activities and the associated exposure to POP contamination makes them suitable species to study regarding the effects of POP contamination in foraging habitats on omnivorous seabirds.

In order to assess the effects of POP contamination on the functioning of gull populations, it is important to pinpoint the different sources of contamination. Since gulls are highly opportunistic and versatile foragers and individuals specialize in certain foraging tactics, individual gulls of the same colony could visit very different foraging habitats, ranging from landfills to the open sea (Camphuysen et al. 2015; Tyson et al. 2015; van Donk et al. 2017). Hence, the source of POP contamination, and thus the degree and nature of the exposure to POPs, is expected to vary greatly between individual gulls. Therefore, to clarify the effect of different POP-contaminated areas on gull populations, it is important to also identify the source of contamination in individual gulls.

Conventional sampling methods applied when studying POP contamination, such as taking liver or fat samples, are destructive and ethically undesired. A less destructive method could be the use of feathers, as it is likely that POPs are deposited in and onto feathers, through, for example, preen oil, blood, or contaminated dust. In fact, feather sampling has been applied for decades to assess the exposure to heavy metals and POPs (Goede and De Voogt 1985; Abbasi et al. 2015). Thus, analyzing differences in POP concentrations in feathers could be a nondestructive way to identify POP sources of individual gulls.

The aim of this literature review was therefore to evaluate the potential of using feathers to determine different sources of POP contamination in individual gulls. This aim was translated into two research questions. The first question was to what extent feathers reflect internal and environmental levels of contamination. Since until now, feather analysis was mainly used to determine the degree of the POP contamination of species inhabiting certain areas, and not to determine where the contamination originated from, the second question was whether it would be possible to distinguish between POP contaminations that originate from different foraging habitats visited by gulls. If this is indeed the case, there are many means to develop similar approaches for studies in other bird species.

2   The Reflection of Internal and Environmental Contaminant Concentrations in Feathers

During the last years, many have studied the possibility to use feathers as a nondestructive biomonitoring tool for persistent organic pollutants (reviewed by García-Fernández et al. 2013). The most commonly studied pollutants are polychlorinated biphenyls (PCBs), organochlorine pesticides (OCPs), and polybrominated diphenyl ethers (PBDEs). These contaminants have been studied all over the world in feathers of a wide variety of bird species, such as predatory birds in Greenland (Jaspers et al. 2011), Norway (Eulaers et al. 2011a, b), Pakistan (Abbasi et al. 2016), Belgium (Jaspers et al. 2006, 2007b; Eulaers et al. 2014), and Argentina (Martínez-López et al. 2015); non-predatory aquatic and terrestrial birds from Iran (Rajaei et al. 2011), the USA (Summers et al. 2010), Belgium (Dauwe et al. 2005; Jaspers et al. 2007b), Spain (Espín et al. 2012), and Romania (Matache et al. 2016); and even poultry in Slovenia (Zupancic-Kralj et al. 1992). Perfluoroalkyl substances (PFAS) have also been detected in feathers of several birds from different trophic levels (Meyer et al. 2009), but the most frequently reported substances are PCBs, OCPs, and PBDEs. Therefore, we will focus on these substances in this section.

Although contaminant concentrations differ among different types of feathers, all studied feather types seem to be adequate biomonitoring tools (reviewed by García-Fernández et al. 2013). There are different pathways of POP deposition into and onto feathers (summarized in Fig. 1). One way is the internal allocation of substances, mainly from the bloodstream, and it has been suggested that this could be a way to sequester harmful substances (Van den Steen et al. 2007). Internal allocation of contaminants to feathers probably occurs during the growth of the feather, when the feather is still connected to the bloodstream (Fig. 1). This implies that especially concentrations of contaminants in newly grown feathers of adult birds and nestlings are related to concentrations in blood and blood plasma (Van den Steen et al. 2007; Eulaers et al. 2011a, b). Concentrations in muscle tissue and fat are also correlated with those in feathers (Dauwe et al. 2005; Jaspers et al. 2006, 2007b; Rajaei et al. 2011; Eulaers et al. 2014). In addition, some studies observed a correlation between POP concentrations in feathers and liver tissue (Rajaei et al. 2011; Eulaers et al. 2014). However, Meyer et al. (2009) only found a correlation between concentrations in feathers and liver tissue when five bird species of different trophic levels were pooled, but not for individual bird species, probably due to a small sample size. Therefore, despite some exceptions, it is concluded that especially newly grown feathers of adult birds and nestlings may reflect the internal contamination profile.

After the feather is fully grown, it is disconnected from the bloodstream, and hence POP concentrations in feathers are less affected by the internal contamination (Fig. 1) (Dauwe et al. 2005). Contamination profiles in fully grown feathers seem to remain rather stable, as it was possible to analyze POP concentrations in feathers of stuffed birds more than 10 years after they were collected (Behrooz et al. 2009).

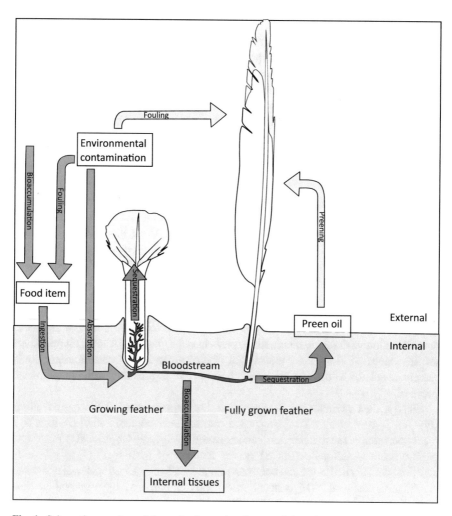

Fig. 1 Schematic overview of the major internal and external deposition pathways of POPs into growing feathers that are connected to the bloodstream (left), and fully grown feathers that are no longer connected to the bloodstream (right). The blue pathways show how environmental contamination enters the bloodstream. POPs can be taken up from the environment directly, via absorption through the skin or in the lungs, or indirectly by ingesting food items that are contaminated by bioaccumulation through the food chain or by fouling with contaminated dust or liquids. In red is the internal pathway that shows the sequestration of POPs from the bloodstream into growing feathers and preen oil. In addition, POPs from the bloodstream can bioaccumulate in internal tissues. In yellow is the external pathway of POP deposition onto feathers that could be by preening with contaminated preen oil and by fouling of the feathers with contaminated dust or liquids

Thus, contamination that is present in the feather at the time of sampling, was probably acquired during feather growth, which can be up to 1 year earlier for flight feathers (Harris 1971). On the contrary, internal POP concentrations could change frequently, as a result of tissue-specific metabolic processes and changed exposure (Jaspers et al. 2006). Therefore, internal body contaminant concentrations represent more recent exposure, and as these concentrations change, the correlation between feather concentrations and internal body concentrations could be weakened in older feathers (Jaspers et al. 2006).

Another pathway of POP deposition on feathers is the external deposition by preening with preen oil (Fig. 1). As older feathers are preened more often than newly grown feathers, this effect changes with the age of the feather (Jaspers et al. 2011). Due to the hydrophobicity of PCBs, PBDEs, and OCPs, preen oil contains relatively high concentrations of POPs (Burreau et al. 2004; Yamashita et al. 2007). Consequently, when preen oil was removed from the feathers, the total POP concentration was significantly reduced in white-tailed eagle (*Haliaeetus albicilla*) and common magpie (*Pica pica*) feathers (Jaspers et al. 2008, 2011). Concentrations of POPs in preen oil correlated with internal POP concentrations in white-tailed eagles (Eulaers et al. 2011a; Jaspers et al. 2011), although this correlation was not observed in water birds by Kocagöz et al. (2014). Nevertheless, preening activity of birds probably enhances the correlation of internal levels of contamination with the contamination levels in older feathers. Hence, after feathers are disconnected from the bloodstream, their contaminant concentrations could remain correlated to the internal concentrations due to preening with contaminated preen oil.

Finally, dust particles could also cause the deposition of POPs onto feathers (Fig. 1). Jaspers et al. (2014) suggested that external contamination by dust at a local point source led to a different ratio between perfluorooctanoic acid (PFOA) and perfluorooctane sulfonate (PFOS) in the feathers of barn owls (*Tyto alba*). In addition, white-tailed eagle and common magpie feathers washed with water showed significantly reduced POP concentrations, possibly due to the removal of dust and preen oil (Jaspers et al. 2008, 2011). Moreover, it has been suggested that contaminated dust on feathers is a source of internal PBDEs in ring-billed gulls (*Larus delawarensis*) that forage on landfills, as a result of dust ingestion when gulls preen their feathers (Gentes et al. 2015). Yet, this seemed of little importance for common buzzards (*Buteo buteo*) and great tits (*Parus major*), possibly due to less exposure to highly polluted dust in their habitats (Dauwe et al. 2005; Jaspers et al. 2007a). The contribution of pollution by dust particles on feathers is probably especially high for adult birds foraging in highly contaminated areas, like landfills. The contribution of this pathway will probably be lower for nestlings, since they are not yet visiting these contaminated areas.

In conclusion, during the growth of the feather, POPs are probably mainly deposited internally, via the bloodstream. Subsequently, when the feather is fully grown and disconnected from the bloodstream, most POPs are probably deposited externally by preening with preen oil and by dust particles (summarized in Fig. 1). These pathways of deposition overlap up to a certain extent, as preen oil is excreted from internal tissues and thus reflects internal contamination levels, and dust particles can be inhaled or ingested and thus also have an effect on the internal levels of contamination. Nevertheless, newly grown feathers reflect recent exposure through internal sequestration and external deposition of contaminated dust particles and preen oil, while older feathers reflect recent exposure through external deposition only.

3 Identification of the Source of Contamination Based on the POP Concentrations in Feathers

One way to distinguish between birds foraging in marine areas or at landfills could be the difference in POP concentrations. Even though the production of PCBs and most PBDEs and OCPs is banned, high concentrations of PCBs and PBDEs are detected in leachate (Öman and Junestedt 2008; Li et al. 2012, 2014) and dust from landfills (Hansen et al. 1997; Melnyk et al. 2015). As shown in Table 1, several studies indicated that concentrations of OCPs, PCBs, and PBDEs in animals are elevated when their habitat is contaminated. However, the source of contamination differs for each POP. OCPs like DDTs were used in European agriculture until the late 1970s and early 1980s (FAO/UNEP 1991), and nowadays elevated concentrations of OCPs are still measured in eggs of great tits in a rural area in Flanders (Van den Steen et al. 2008) (Table 1). Also in common magpie feathers from Flanders, DDE concentrations were higher in rural samples compared to urban samples (Jaspers et al. 2009) (Table 1). Therefore, high concentrations of OCPs in feathers could indicate a rural foraging area. However, several raptors collected in a Chinese urban area contained high DDT concentrations, up to 158,700 ng g^{-1} DDT in Eurasian sparrowhawks (*Accipiter nisus*), that could be the result of highly contaminated wintering or stopover sites in southeast China (Chen et al. 2009). Although the use of DDT in China has been banned in 1983, no apparent decline in DDT concentrations has been observed in the field, and large amounts of DDT are still produced and probably discharged as a result of export demands and the production of dicofol (Qiu et al. 2005; Zhao et al. 2018).

Table 1 shows that, in contrast to OCPs, elevated concentrations of PCBs and PBDEs in birds were mostly linked to urban areas (Jaspers et al. 2009; François et al. 2016; Zeng et al. 2016), industry (Batty et al. 1990; Smith et al. 2003; Van den Steen et al. 2008; Zeng et al. 2016), and landfills (Johnson et al. 1996; Halbrook and Arenal 2003; Chen et al. 2013; Gilchrist et al. 2014), and the study of Van den Steen et al. (2008) showed that PCB and PBDE concentrations were highly correlated to each other. Moreover, Ito et al. (2013) measured elevated PCB concentrations in preen oil of GPS-tracked streaked shearwaters (*Calonectris leucomelas*) that foraged in an inland sea surrounded by urbanized coast, compared to shearwaters that foraged in the Pacific Ocean. On the contrary, De la Casa-Resino et al. (2015) did not measure any detectable concentrations of PCBs in white stork chicks (*Ciconia*

Table 1 Overview of reported mean or median DDT, PCB, and PBDE concentrations (ng g^{-1}) detected in samples collected from birds inhabiting different areas

POP	Species	Animal	Landfill (ng g^{-1})	Urban (ng g^{-1})	Industry (ng g^{-1})	Rural (ng g^{-1})	Sample	Median/mean	Reference
Σ DDTs	Pica pica	Common magpie	–	3.07	–	34.2	Feathers	Median	(Jaspers et al. 2009)
	Parus major	Great tit	–	–	400–1100	500–2700	Eggs	Median	(Van den Steen et al. 2008)
	Falco tinnunculus	Common kestrel	–	8600	–	–	Liver	Median	(Chen et al. 2009)
	Athene noctua	Little owl	–	23,200	–	–	Liver	Median	(Chen et al. 2009)
	Otus sunia	Scops owl	–	16,900	–	–	Liver	Median	(Chen et al. 2009)
	Asio otus	Long-eared owl	–	1100	–	–	Liver	Median	(Chen et al. 2009)
	Accipiter nisus	Eurasian sparrowhawk	–	158,700	–	–	Liver	Median	(Chen et al. 2009)
	Accipiter gularis	Japanese sparrowhawk	–	62,700	–	–	Liver	Median	(Chen et al. 2009)
	Buteo buteo and Buteo hemilasius	Common and upland buzzard	–	1900	–	–	Liver	Median	(Chen et al. 2009)
*	Ciconia ciconia	White stork	ND	–	–	0.11	Nestling blood	Median	(de la Casa-Resino et al. 2015)
Σ PCBs	Pica pica	Common magpie	–	140	–	4.24	Feathers	Median	(Jaspers et al. 2009)
	Corvus macrorhynchos	Jungle crow	540	–	–	2100	Breast muscle	Mean	(Watanabe et al. 2005)
*	Corvus splendens	House crow	1200	–	–	820	Breast muscle	Mean	(Watanabe et al. 2005)
	Parus major	Great tit	–	–	2000–6000	1000–3700	Eggs	Median	(Van den Steen et al. 2008)
	Sturnus vulgaris	European starling	4100–27,700	–	–	300	Eggs	Mean	(Halbrook and Arenal 2003)
	Falco tinnunculus	Common kestrel	–	3500	–	–	Liver	Median	(Chen et al. 2009)
	Accipiter nisus	Eurasian sparrowhawk	–	19,900	–	–	Liver	Median	(Chen et al. 2009)
	Accipiter gularis	Japanese sparrowhawk	–	3900	–	–	Liver	Median	(Chen et al. 2009)

	Species	Common name					Sample type		Reference
	Buteo buteo and *Buteo hemilasius*	Common and upland buzzard	–	200	–	–	Liver	Median	(Chen et al. 2009)
	Athene noctua	Little owl	–	1500	–	–	Liver	Median	(Chen et al. 2009)
	Otus sunia	Scops owl	–	500	–	–	Liver	Median	(Chen et al. 2009)
	Asio otus	Long-eared owl	–	100	–	–	Liver	Median	(Chen et al. 2009)
	Gallinago gallinago	Common snipe	–	–	2200	–	Pectoral muscle	Median	(Luo et al. 2009)
*	*Ciconia ciconia*	White stork	ND	–	–	ND	Nestling blood	Median	(de la Casa-Resino et al. 2015)
	Ardeola bacchus	Chinese-pond heron	–	–	2200	–	Pectoral muscle	Median	(Luo et al. 2009)
	Calonectris leucomelas	Streaked shearwaters	–	1600	–	200	Preen oil	Median	(Ito et al. 2013)
	Amaurornis phoenicurus	White-breasted waterhen	–	–	600	–	Pectoral muscle	Median	(Luo et al. 2009)
	Gallirallus striatus	Slaty-breasted rail	–	–	820	–	Pectoral muscle	Median	(Luo et al. 2009)
	Porzana fusca	Ruddy-breasted crake	–	–	37	–	Pectoral muscle	Median	(Luo et al. 2009)
Σ PBDEs	*Pica pica*	Common magpie	–	0.41	–	0.27	Feathers	Median	(Jaspers et al. 2009)
	Parus major	Great tit	–	–	55–80	35–60	Eggs	Median	(Van den Steen et al. 2008)
	Sturnus vulgaris	European starling	30–280	–	15–102	6.7–44	Eggs	Median	(Chen et al. 2013)
	Tachycineta bicolor	Tree swallow	–	205.5–590.1	–	83.6	Eggs	Mean	(Gilchrist et al. 2014)
	Larus delawarensis	Ring-billed gull	–	128	–	–	Liver	Mean	(François et al. 2016)
	Gallus gallus	Chicken	–	–	7700	3100	Eggs	Median	(Zeng et al. 2016)
*	*Anser anser*	Goose	–	700	–	400	Eggs	Median	(Zeng et al. 2016)

Areas are divided into four categories: landfills, urban areas, industrial areas, and rural areas. Also the sample type and whether the provided value is a median or a mean are provided. Within chemical compounds, species are ordered taxonomically (Laurin and Gauthier 2012). Concentrations specified with ND (no data) were analyzed, but were below detection limit. Concentrations specified with a dash (–) were not analyzed. Rows marked with an asterisk (*) indicate that contaminant concentrations at different locations did not differ significantly ($p > 0.05$)

ciconia) in a nest close to a landfill, even though white storks in this rural area visited the landfill frequently (Table 1). In addition, magpie feathers from urban, rural, and industrial sites did not exhibit different concentrations of PBDEs (Table 1) (Jaspers et al. 2009). Watanabe et al. (2005) did not observe a significantly different PCB concentration in the breast muscles of house crows (*Corvus splendens*) living on an Indian landfill compared to rural house crows, while the PCB concentration in the breast muscles of rural jungle crows (*Corvus macrorhynchos*) was even significantly higher than in crows on landfills. However, both crow species from landfills exhibited significantly higher concentrations of the more harmful dioxin-like PCBs (Watanabe et al. 2005). Finally, no significantly different PBDE concentrations were reported for urban domestic goose eggs (*Anser anser*), although concentrations appeared to be higher (Zeng et al. 2016). Nevertheless, as is shown in (Table 1), there is substantial evidence in the literature that elevated PCB and PBDE concentrations in birds can be linked to urban or industrial areas and landfills.

Therefore, gulls that forage in PCB-, PBDE-, or OCP-contaminated areas are likely to contain elevated concentrations of these contaminants in their body, eggs, and feathers. In this regard, PCBs and PBDEs could be especially useful, since elevated concentrations of these POPs are linked to landfills and urban and industrialized areas. The next step would therefore be to evaluate if specific PCB and PBDE profiles could give information about specific POP sources.

4 Identification of the Source of Contamination Based on the POP Congener Profile in Feathers

4.1 Linking the POP Congener Profile to the Source of Contamination

The POP congener composition could also provide valuable information regarding the source of contamination in gulls. In total, there are 209 PCB and 209 PBDE congeners, numbered after the position and number of chlorine (in PCBs) or bromine (in PBDEs) atoms [Fig. 2: PCB (a) and PBDE (b) molecules. The numbers indicate the positions that can be halogenated with chlorine (PCB) or bromine (PBDE) atoms]. The degree of biomagnification between trophic levels is determined by the bioavailability, the uptake, the excretion, and the ability of the animal to metabolize or dehalogenate the congener (biotransformation) (Arnot and Gobas 2003; Burreau et al. 2004). These factors depend greatly on the halogenation of the congener and the physiology and metabolic capacity of the animal (Hawker and Connell 1988; Boon et al. 1989, 1994; Fisk et al. 1999; Arnot and Gobas 2003; Voorspoels et al. 2007). Highly chlorinated PCBs (≥ 6 chlorines) are more hydrophobic than the lightly chlorinated PCBs (≤ 5 chlorines) and are also metabolized more slowly than the lightly chlorinated (<5 chlorines) compounds (Boon et al. 1989), making them highly bioaccumulative (Arnot and Gobas

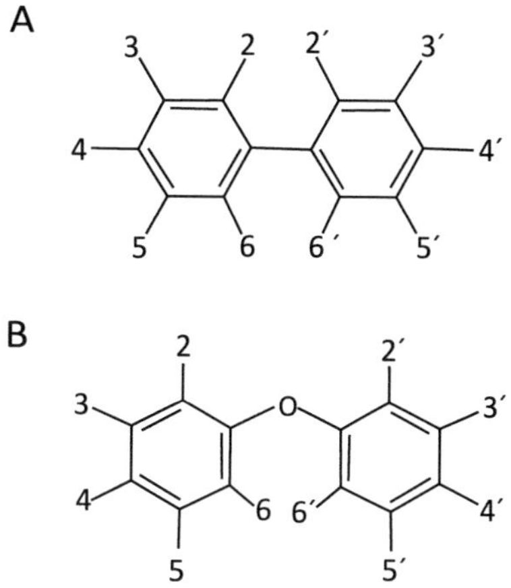

Fig. 2 PCB (**a**) and PBDE
(**b**) molecules. Numbers
indicate positions that can
be halogenated with
chlorine (PCB) or bromine
(PBDE) atoms

2003; Burreau et al. 2004). In contrast to PCBs, for PBDEs especially the lightly brominated congeners (≤ 5 bromines) bioaccumulate strongly, but the bioaccumulation of highly brominated PBDEs (≥ 6 bromines) is restricted by their slow uptake due to their large size and high molecular weight and their metabolic debromination after uptake (de Wit 2002; Burreau et al. 2004, 2006; Van den Steen et al. 2007; Voorspoels et al. 2007; Letcher et al. 2014; François et al. 2016). Therefore, highly chlorinated (≥ 6 Cl) PCBs and lightly brominated (≤ 5 Br) PBDEs have a comparably high biomagnification potential, in contrast to lightly (≤ 5 Cl) chlorinated PCB and highly (≥ 6 Br) brominated PBDE congeners (Burreau et al. 2004).

Congener-specific biomagnification rates and site-specific contamination sources are likely to result in different PCB and PBDE congener profiles. A gull foraging at sea is mainly exposed to POPs through the bioaccumulation of POPs in its prey, and hence the trophic position of the gull and the prey plays an important role in the exhibited congener composition in the body and feathers (Borgå et al. 2001; Ruus et al. 2002). A higher proportion of more bioaccumulative congeners will probably occur in these gulls due to biomagnification and metabolism via the food chain (Strandberg et al. 1998; Dietz et al. 2000; Borgå et al. 2005). As we explained in the section above, highly chlorinated PCBs and lightly brominated PBDEs are more bioaccumulative, and, therefore, higher proportions of these congeners are likely to be present in gulls foraging at sea.

In contrast to gulls foraging at sea, gulls foraging at landfills predominantly feed on anthropogenic food. The food itself has in general relatively low POP concentrations but can be covered by leachate and dust containing high POP concentrations (Brousseau et al. 1996; Duhem et al. 2003, 2005; Schecter et al. 2010; Huwe and Larsen 2005; Li et al. 2012; McFarland and Clarke 1989; Öman and Junestedt 2008; Hansen et al. 1997; Persson et al. 2005). Gulls are thus probably mainly exposed to POPs by eating food items or preening feathers that are fouled with contaminated dust or leachate (Persson et al. 2005; Gentes et al. 2015). In addition, substantial amounts of POPs could be absorbed when lungs, skin, or feathers are regularly exposed to contaminated dust, leachate, and aerosols. Due to the relatively high proportion of less bioaccumulative congeners in these substances, gulls that foraged on landfills are likely exposed to a higher proportion of less bioaccumulative congeners, which could be reflected by the congener profile of their feathers (Hansen et al. 1997; Öman and Junestedt 2008; Melnyk et al. 2015). A higher proportion of less bioaccumulative congeners in birds inhabiting contaminated areas is supported by several studies. European starlings (*Sturnus vulgaris*) nesting on a landfill and common magpies living in urban areas exhibited higher proportions of lightly chlorinated and thus less bioaccumulative PCBs in their eggs or feathers, compared to starlings or magpies living in a less contaminated area (Halbrook and Arenal 2003; Jaspers et al. 2009). In addition, a high proportion and concentration of the fully brominated and therefore less bioaccumulative decabromodiphenylether (DecaBDE or BDE 209) was found in 25% of male ring-billed gulls that visited refuse tips at least once (Gentes et al. 2015) and elevated concentrations of highly brominated PBDEs were found in eggs of great tits and tissue of ring-billed gulls that inhabit urban areas (Van den Steen et al. 2008; François et al. 2016). Nevertheless, in contrast to what would be expected based on the level of bromination, the tetrabrominated PBDE 47 was more prominent in urban common magpie feathers, while the pentabrominated PBDE 99 was more prominent in rural magpie feathers (Jaspers et al. 2009). However, despite this last exception, a higher proportion of less bioaccumulative congeners is in general exhibited in birds inhabiting contaminated areas.

Thus, based on the combined evidence described in this section, we conclude that the analysis of the congener profiles in gull feathers could be a promising approach to determine the likely source of contamination in gulls. The trophic position of gulls foraging at sea will likely cause a higher proportion of more bioaccumulative POPs, such as highly chlorinated PCBs and lightly brominated PBDEs. Gulls that forage on landfills will probably exhibit a higher proportion of less bioaccumulative congeners, due to a relatively high availability of these congeners in these areas. This approach will be further demonstrated by means of a case study in the next section.

4.2   Case Study: The [PCB 153]/[PCB 52] and [PCB 118]/ [PCB 52]-Ratios

To demonstrate the feasibility of the analysis of differences in congener profiles to assess the contamination source, we performed a case study regarding the ratio between the concentrations of the highly bioaccumulative PCB 153 and PCB 118 (2,2′,4,4′,5,5′-hexachlorobiphenyl and 2,3′,4,4′,5-pentachlorobiphenyl, respectively) and the less bioaccumulative PCB 52 (2,2′,5,5′-tetrachlorobiphenyl) (Borgå et al. 2004). Based on the theory explained in the previous paragraph, we hypothesized that relatively low ratios are exhibited in birds that foraged on landfills and in urban areas, and relatively high ratios are exhibited in birds that foraged at sea.

To test this hypothesis, we calculated the [PCB 118]/[PCB 52] and [PCB 153]/[PCB 52] ratios for animals inhabiting natural areas or urban areas and landfills, from concentrations obtained from studies that measured PCB concentrations in feather, liver, and preen oil/gland samples in birds (references in Table 3). In addition, to assess the bioavailability of these congeners at landfills, we calculated the ratio from concentrations obtained from studies that measured PCBs in different landfill samples (references in Table 2). The results from this calculation are summarized in Tables 2 and 3 and Fig. 3.

As we hypothesized, relatively high ratios were mostly obtained for feather, preen oil, and liver samples from birds that foraged in a natural environment (Fig. 3 and Table 3). This is presumably due to a higher degree of biomagnification of the more bioaccumulative PCBs 118 and 153 in their prey, which is indicated by the high ratios obtained for fish and crustaceans in natural environments (Table 3) (Duhem et al. 2005; Abdennadher et al. 2014). Simultaneously, usually low ratios were obtained for feather, liver, and preen gland samples from birds that forage in urban areas or at landfills, which is presumably due to the availability of similar proportions of PCBs 52, 118, and 153 in these areas (Tables 2 and 3 and Fig. 3). Indeed, the low ratios obtained in different landfill samples indicate a similar or even higher availability of PCB 52 compared to PCBs 118 and 153 (Table 2).

The ratios in liver tissue in particular for birds from natural areas were higher than the ratios for birds from landfills (Fig. 3). This pattern was also generally observed in feathers and preen gland tissue/oil, although there were some exceptions (Fig. 3). First of all, relatively low ratios were calculated for feathers of white-tailed eagles from a natural environment, compared to the high [PCB 153]/[PCB 52] ratio obtained from preen oil of the same species (Table 3, Fig. 3) and to the ratios that were obtained from feathers of common magpies from an industrial urban area (Table 3, Fig. 3). Furthermore, a relatively high [PCB 153]/[PCB 52] ratio was obtained from the preen gland of common magpies inhabiting an industrial urban area (Flanders, Belgium), especially compared to the [PCB 118]/[PCB 52] ratio in the preen gland from the same individuals. This might be due to a local source of PCB 153 (Table 3, Fig. 3) (Jaspers et al. 2008). However, no such high [PCB 153]/[PCB 52] ratio was obtained from common magpie feathers from the same area (Table 3, Fig. 3) (Jaspers et al. 2008).

Table 2 [PCB 153]/[PCB 52] and [PCB 118]/[PCB 52] ratios in different landfill samples, calculated from concentrations obtained from different studies

Sample	Country	[PCB 153]/[PCB 52]	[PCB 118]/[PCB 52]	Reference
Leachate sediment	Canada	1.33	0.56	(Öman and Junestedt 2008)
Dust	USA	0.98	–	(Hansen et al. 1997)
Surface soil	Poland	0.68	0.75	(Melnyk et al. 2015)

Table 3 [PCB 153]/[PCB 52] and [PCB 118]/[PCB 52] ratios for different species from different trophic levels and habitats, calculated from concentrations obtained from different studies

Species	Tissue	Trophic level	Area	[PCB 153]/[PCB 52]	[PCB 118]/[PCB 52]	Diet	Reference
Zooplankton	Homogenized	Primary/secondary consumer	Natural area, northern Baltic Sea	3.25	1.06	Phytoplankton, zooplankton, detritus	(Strandberg et al. 1998)
Zooplankton	Homogenized	Primary/secondary consumer	Natural area, mid Baltic Sea	1.36	1.23	Phytoplankton, zooplankton, detritus	(Strandberg et al. 1998)
Mysis sp.	Homogenized	Secondary consumer	Natural area, northern Baltic Sea	6.62	3.69	Zooplankton	(Strandberg et al. 1998)
Mysis sp.	Homogenized	Secondary consumer	Natural area, mid Baltic Sea	4.15	2.31	Zooplankton	(Strandberg et al. 1998)
Clupea harengus	Homogenized	Secondary consumer	Natural area, northern Baltic Sea	7.33	3.17	Crustaceans, zooplankton, fish	(Strandberg et al. 1998)
Clupea harengus	Homogenized	Secondary consumer	Natural area, mid Baltic Sea	9.17	4.58	Crustaceans, zooplankton, fish	(Strandberg et al. 1998)
Cepphus grylle	Liver	Omnivore	Natural area, Barents Sea	12.5	8.38	Fish, crustaceans, and molluscs, insects, plants	(Borgå et al. 2005)
Uria lomvia	Liver	Secondary consumer	Natural area, Barents Sea	25	22.5	Fish, crustaceans, and molluscs	(Borgå et al. 2005)
Haliaeetus albicilla	Feathers	Top predator	Natural area, Trondelag, NO	6.88	2.8	Fish, birds, and mammals	(Eulaers et al. 2011b)
Haliaeetus albicilla	Preen oil	Top predator	Natural area, Trondelag, NO	31	8	Fish, birds, and mammals	(Eulaers et al. 2011b)
Larus audouinii	Eggs	Secondary consumer	Mediterranean Sea	–	3.78–25	Fish	(Goutner et al. 2001)
Sturnus vulgaris	Chicks	Omnivore	Landfill, Illinois, USA	2.5	2.55	Invertebrates, seeds, fruit	(Halbrook and Arenal 2003)
Sturnus vulgaris	Chicks	Omnivore	Landfill, Illinois, USA	4.18	3.5	Invertebrates, seeds, fruit	(Halbrook and Arenal 2003)
Pica pica	Feathers	Omnivore	Urban area, Antwerp, BE	2.59	3.89	Young birds and eggs, small mammals, insects, scraps, acorns	(Jaspers et al. 2008)
Pica pica	Preen gland	Omnivore	Urban area, Antwerp, BE	48	3.4	Young birds and eggs, small mammals, insects, scraps, acorns	(Jaspers et al. 2008)

Sample type, animal type, and a brief diet overview are also provided

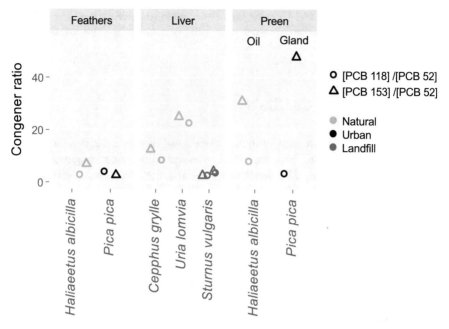

Fig. 3 [PCB 153]/[PCB 52] ratio (circles) and [PCB 118]/[PCB 52] ratio (triangles) for from either natural habitats (yellow) or urban areas (black) and landfills (red). Please note that preen oil and preen gland samples from *Haliaeetus albicilla* and *Pica pica*, respectively, are compared. Exact values, a more detailed description of the data and references can be found in Table 3

These exceptions could be due to species- and tissue-specific characteristics. Most likely, the higher chlorinated congeners have a higher affinity for preen oil than for feathers. These observations clearly indicate that preen oil PCB concentrations cannot be simply compared to concentrations and ratios in feathers, and when analyzing feathers, the influence of the congener profile of preen oil spread on the feathers should be taken into account.

Nevertheless, despite some exceptions, the ratios calculated for birds that inhabit natural areas, especially those for different seabird species, were generally substantially higher than the ratios for birds that foraged on landfills or in urban areas. Therefore, the analysis of POP congener profile in feathers and calculation of the ratios between more and less-accumulative congeners could be a promising approach to determine the source of contamination in gulls and is worth further investigation. Further species-specific analysis of a wide variety of PCB and PBDE congeners in feathers, combined with the analysis of PCB and PBDE congers in leachate, dust, and surface soil samples from landfills visited by the birds, according to Watanabe et al. (2005) and Table 2, could provide a stronger empirical basis for this approach. In addition, combining congener profile analysis in feathers with the analysis of stable isotopes signatures (Hobson and Clark 1992; Hobson 1993; Moreno et al. 2010; Auman et al. 2011; Caron-Beaudoin et al. 2013) on a subsample from the same feathers and GPS tagging of the birds to quantify the time spent in different habitats (e.g., Camphuysen et al. 2015) could make this method very powerful.

5 Discussion and Conclusions

Based on the literature we studied for this review, we conclude that it is most likely possible to distinguish between POP contamination that originates from different foraging areas, like landfills or marine environments, based on the congener concentrations and profiles in gull feathers. Environmental and internal concentrations were to a certain extent reflected by the concentrations in feathers of adult birds and nestlings. In addition, it is likely possible to distinguish between different foraging habitats by combining the analysis of the total POP concentrations with the determination of the ratios between more-accumulative and less-accumulative PCB and PBDE congeners. However, this conclusion was drawn from the combined evidence of different studies, concerning a wide variety of species and tissues. Although PCB and PBDE concentrations in feathers are to a certain extent related to internal tissues, caution is necessary when comparing different tissues. Therefore, more insight is required into the establishment of POP congener concentrations and profiles in feathers in relation to the source of POP contamination, before using this approach.

Several aspects of this approach should be taken into account. First of all, the age of the birds greatly affects the exposure, since nestlings are unlikely to come in direct contact with environmental contamination, and young adults need to develop a specialization. Moreover, the type and age of the feather determine to what extent the POP concentrations in the feather reflect recent exposure and internal contaminant concentrations. Hence, especially newly grown feathers of adults are very suitable for this analysis, as they reflect both recent exposure through internal sequestration of contaminants and external deposition of dust particles from contaminated foraging areas. Sampling shed feathers is not possible for this approach, because external contamination might be worn off. Secondly, individual specializations in foraging strategies can lead to a large variety of foraging habitats and degrees of POP exposure, due to the large diversity in POP bioavailability between foraging habitats and even specific foraging locations. Furthermore, preening, washing, and swimming also affect the outer POP concentration, and thus influence to what extent POP concentrations on the feather reflect environmental and internal contamination. Finally, the concentrations of PCB and PBDE congeners in feathers are determined by the local bioavailability and the chemical properties of the congeners. These complications should be taken into account when analyzing POPs in gull feathers to identify contamination source.

In order to gain more insight into the complications of POP analysis in gull feathers and to provide a stronger basis to implement this approach, more research is required. We advise to further investigate the differences in POP concentrations between recently grown and older feathers from adult or preadult gulls and to study the concentrations and composition of a wide variety of PCB and PBDE congeners inside the feather as well as on the outer surface, for example, conform the method of Jaspers et al. (2008). We also advise investigating the relation between the POP concentrations and congener compositions in preen oil and the concentrations and congener profiles in and on the feathers.

A case study of Jaspers et al. (2014) showed that the analysis of POP signatures in animal tissue may identify a specific point source of contamination, in this case the manufacturer of these POPs. However, when the possible source of contamination is a landfill, the larger variety of POPs that originate from this source complicates the analysis (Hansen et al. 1997; Öman and Junestedt 2008; Melnyk et al. 2015). Therefore, POP analysis could be combined with the analysis of stable isotopic signatures and GPS tracking. The identification of foraging habitats by tracking bird movements with bird-borne GPS loggers could be a crucial step in directly linking individual contamination profiles to a contamination source (Ito et al. 2013; Gentes et al. 2015). In addition, the analysis of stable isotopic signatures of carbon, nitrogen, and sulfur isotopes in feathers can provide information regarding the foraging area, trophic level, and diet composition (Hobson and Clark 1992; Hobson 1993; Moreno et al. 2010; Auman et al. 2011; Caron-Beaudoin et al. 2013). This would allow for an evaluation of the strength of the relationship between the sources of contamination obtained from POP analysis and the foraging habitats derived from GPS tracking and the analysis of stable isotopic signatures. For example, in white storks breeding in the vicinity of a landfill (1.5 and 4.9 km) in Spain, no PCBs but high concentrations of DDTs were detected (de la Casa-Resino et al. 2015). This could indicate that these birds forage in agricultural areas rather than on landfills, despite the close proximity of the landfill to their nests. GPS tracking, to determine their actual habitat use, and stable isotope analysis, to determine their trophic level and distinguish direct POP exposure in landfills from POP uptake via the food chain, would provide important complementary information (Abdennadher et al. 2014; Sommerfeld et al. 2016). Gentes et al. (2015) successfully related individual contamination in gulls to foraging habitat use, by combining GPS tracking with the analysis of PBDE concentrations in blood plasma. However, a strong link between habitat use and contamination in and on feathers instead of in blood plasma or certain tissues has not yet been made. This could be a crucial step in testing and further develop the proposed nondestructive approach. Moreover, when fully developed, this approach could be applied to many more individuals and on a far broader scale than would ever be possible when using GPS tags.

Finally, in this review gulls were the focal species, but the analysis of POPs in feathers could probably also be applied when studying other bird species. However, other bird species might have a totally different ecology that induces complications that are not discussed in this review. For example, not all bird species produce preen oil to treat their feathers, some species use powder down or do not use any substance for preening (Wetmore 1920; Kenyon Ross 1976). This will almost certainly affect the POP concentrations in and on their feathers and therefore should be taken into account.

In conclusion, despite some uncertainties that might be reduced by future research, enough evidence was obtained from the reviewed literature to propose the analysis of POPs in newly grown feathers of adult gulls and nestlings as a promising nondestructive approach to analyze the exposure of gulls to POPs and to identify the source of contamination. It could probably be extended to analyze sources of POP contamination in other bird species, provided that complications regarding the biology of the species are taken into account. Especially when integrated with other methods, like GPS tagging and stable isotope analysis, our proposed approach could prove to be very powerful.

6 Summary

Persistent organic pollutants (POPs) are present in almost all environments due to their high bioaccumulation potential. Especially species that adapted to human activities, like gulls, might be exposed to harmful concentrations of these chemicals. The nature and degree of the exposure to POPs greatly vary between individual gulls, due to their diverse foraging behavior and specialization in certain foraging tactics. Therefore, in order clarify the effect of POP-contaminated areas on gull populations, it is important to identify the sources of POP contamination in individual gulls. Conventional sampling methods applied when studying POP contamination are destructive and ethically undesired. The aim of this literature review was to evaluate the potential of using feathers as a nondestructive method to determine sources of POP contamination in individual gulls. The reviewed data showed that high concentrations of PCBs and PBDEs in feathers together with a large proportion of less bioaccumulative congeners may indicate that the contamination originates from landfills. Low PCB and PBDE concentrations in feathers and a large proportion of more bioaccumulative congeners could indicate that the contamination originates from marine prey. We propose a nondestructive approach to identify the source of contamination in individual gulls based on individual contamination levels and PCB and PBDE congener profiles in feathers. Despite some uncertainties that might be reduced by future research, we conclude that especially when integrated with other methods like GPS tracking and the analysis of stable isotopic signatures, identifying the source of POP contamination based on congener profiles in feathers could become a powerful nondestructive method.

Acknowledgments We would like to thank the anonymous reviewer for the useful comments that helped improve the manuscript.

References

Abbasi NA, Khan MU, Jaspers VLB et al (2015) Spatial and interspecific variation of accumulated trace metals between remote and urbane dwelling birds of Pakistan. Ecotoxicol Environ Safe 113:279–286. https://doi.org/10.1016/j.ecoenv.2014.11.034

Abbasi NA, Eulaers I, Jaspers VLB et al (2016) Use of feathers to assess polychlorinated biphenyl and organochlorine pesticide exposure in top predatory bird species of Pakistan. Sci Total Environ 569–570:1408–1417. https://doi.org/10.1016/j.scitotenv.2016.06.224

Abdennadher A, Ramírez F, Romdhane MS et al (2014) Using a three-isotope Bayesian mixing model to assess the contribution of refuse dumps in the diet of yellow-legged gull Larus michahellis. Ardeola 61:297–309. https://doi.org/10.13157/arla.61.2.2014.297

Arizaga J, Aldalur A, Herrero A et al (2014) Foraging distances of a resident yellow-legged gull (Larus michahellis) population in relation to refuse management on a local scale. Eur J Wildl Res 60:171–175. https://doi.org/10.1007/s10344-013-0761-4

Arnot JA, Gobas FAPC (2003) A generic QSAR for assessing the bioaccumulation potential of organic chemicals in aquatic food webs. Qsar Comb Sci 22:337–345. https://doi.org/10.1002/qsar.200390023

Auman HJ, Bond AL, Meathrel CE, Richardson AMM (2011) Urbanization of the silver gull: evidence of anthropogenic feeding regimes from stable isotope analyses. Waterbirds 34:70–76. https://doi.org/10.1675/063.034.0108

Batty J, Leavitt RA, Biondot N, Polin D (1990) An ecotoxicological study of a population of the white footed mouse (Peromyscus leucopus) inhabiting a polychlorinated biphenyls-contaminated area. Arch Environ Contam Toxicol 19:283–290

Baxter AT, Allan JR (2006) Use of raptors to reduce scavenging bird numbers at landfill sites. Wildl Soc Bull 34:1162–1168. https://doi.org/10.2193/0091-7648(2006)34[1162:uortrs]2.0.co;2

Behrooz RD, Esmaili-Sari A, Ghasempouri SM et al (2009) Organochlorine pesticide and polychlorinated biphenyl residues in feathers of birds from different trophic levels of South-West Iran. Environ Int 35:285–290. https://doi.org/10.1016/j.envint.2008.07.001

Belant JL (1997) Gulls in urban environments: landscape-level management to reduce conflict. Landsc Urban Plan 38:245–258. https://doi.org/10.1016/S0169-2046(97)00037-6

Belant JL, Seamans TW, Gabrey SW, Ickes SK (1993) Importance of a landfill to nesting herring gulls. Condor 95:817–830

Belant JL, Ickes SK, Seamans TW (1998) Importance of landfills to urban-nesting herring and ring-billed gulls. Landsc Urban Plan 43:11–19. https://doi.org/10.1016/S0169-2046(98)00100-5

Bernhardt GE, Blackwell BF, Devault TL, Kutschbach-Brohl L (2010) Fatal injuries to birds from collisions with aircraft reveal anti-predator behaviours. Ibis 152:830–834. https://doi.org/10.1111/j.1474-919X.2010.01043.x

Bicknell AWJ, Oro D, Camphuysen KCJ, Votier SC (2013) Potential consequences of discard reform for seabird communities. J Appl Ecol 50:649–658. https://doi.org/10.1111/1365-2664.12072

Blanco G (1996) Population dynamics and communal roosting of White Storks foraging at a Spanish refuse dump. Colon Waterbirds 19:273–276

Boon JP, Eijgenraam F, Everaarts JM, Duinker JC (1989) A structure-activity relationship (SAR) approach towards metabolism of PCBs in marine animals from different trophic levels. Mar Environ Res 27:159–176. https://doi.org/10.1016/0141-1136(89)90022-6

Boon JP, Oostingh I, van der Meer J, Hillebrand MTJ (1994) A model for the bioaccumulation of chlorobiphenyl congeners in marine mammals. Eur J Pharmacol Environ Toxicol 270:237–251. https://doi.org/10.1016/0926-6917(94)90068-X

Borgå K, Gabrielsen GW, Skaare JU (2001) Biomagnification of organochlorines along a Barents Sea food chain. Environ Pollut 113:187–198. https://doi.org/10.1016/S0269-7491(00)00171-8

Borgå K, Fisk AT, Hoekstra PE, Muir DCG (2004) Biological and chemical factors of importance in the bioaccumulation and trophic transfer of persistent organochlorine contaminants in Arctic marine food webs. Environ Toxicol Chem 23:2367–2385. https://doi.org/10.1897/03-518

Borgå K, Wolkers H, Skaare JU et al (2005) Bioaccumulation of PCBs in Arctic seabirds: influence of dietary exposure and congener biotransformation. Environ Pollut 134:397–409. https://doi.org/10.1016/j.envpol.2004.09.016

Bosch M, Oro D, Ruiz X (1994) Dependence of yellow-legged gulls (Larus cachinnans) on food from human activity in two Western Mediterranean colonies. Avocetta 18:135–139

Brousseau AP, Lefebvre J, Giroux J, Waterbirds SC (1996) Diet of Ring-billed Gull chicks in urban and non-urban colonies in Quebec. Waterbird Soc 19:22–30

Burreau S, Zebühr Y, Broman D, Ishaq R (2004) Biomagnification of polychlorinated biphenyls (PCBs) and polybrominated diphenyl ethers (PBDEs) studied in pike (Esox lucius), perch (Perca fluviatilis) and roach (Rutilus rutilus) from the Baltic Sea. Chemosphere 55:1043–1052. https://doi.org/10.1016/j.chemosphere.2003.12.018

Camphuysen KCJ, Shamoun-Baranes J, Van Loon EE, Bouten W (2015) Sexually distinct foraging strategies in an omnivorous seabird. Mar Biol 162:1417–1428. https://doi.org/10.1007/s00227-015-2678-9

Caron-Beaudoin É, Gentes ML, Patenaude-Monette M et al (2013) Combined usage of stable isotopes and GPS-based telemetry to understand the feeding ecology of an omnivorous bird, the Ring-billed Gull (Larus delawarensis). Can J Zool 91:689–697. https://doi.org/10.1675/1524-4695(2008)31[122:SMDAFC]2.0.CO;2

de la Casa-Resino I, Hernández-Moreno D, Castellano A et al (2014) Breeding near a landfill may influence blood metals (Cd, Pb, Hg, Fe, Zn) and metalloids (Se, As) in white stork (Ciconia ciconia) nestlings. Ecotoxicology 23:1377–1386. https://doi.org/10.1007/s10646-014-1280-0

de la Casa-Resino I, Hernández-Moreno D, Castellano A et al (2015) Chlorinated pollutants in blood of White stork nestlings (Ciconia ciconia) in different colonies in Spain. Chemosphere 118:367–372. https://doi.org/10.1016/j.chemosphere.2014.10.062

de Wit CA (2002) An overview of brominated flame retardants in the environment. Chemosphere 46:583–624. https://doi.org/10.1016/S0045-6535(01)00225-9

Chen D, Zhang X, Mai B et al (2009) Polychlorinated biphenyls and organochlorine pesticides in various bird species from northern China. Environ Pollut 157:2023–2029. https://doi.org/10.1016/j.envpol.2009.02.027

Chen D, Martin P, Burgess NM et al (2013) European starlings (Sturnus vulgaris) suggest that landfills are an important source of bioaccumulative flame retardants to Canadian terrestrial ecosystems. Environ Sci Technol 47(21):12238–12247

Christel I, Navarro J, del Castillo M et al (2012) Foraging movements of Audouin's gull (Larus audouinii) in the Ebro Delta, NW Mediterranean: a preliminary satellite-tracking study. Estuar Coast Shelf Sci 96:257–261. https://doi.org/10.1016/j.ecss.2011.11.019

Dauwe T, Jaspers VLB, Covaci A et al (2005) Feathers as a nondestructive biomonitor for persistent organic pollutants. Environ Toxicol Chem 24:442. https://doi.org/10.1897/03-596.1

Dietz R, Riget F, Cleemann M et al (2000) Comparison of contaminants from different trophic levels and ecosystems. Sci Total Environ 245:221–231. https://doi.org/10.1016/S0048-9697(99)00447-7

Dolbeer RA, Belant JL, Sillings JL (1993) Shooting gulls reduces strikes with aircraft at John F. Kennedy International Airport. Wildl Soc Bull 21:442–450

van Donk S, Camphuysen KCJ, Shamoun-Baranes J, van der Meer J (2017) The most common diet results in low reproduction in a generalist seabird. Ecol Evol 4620–4629. https://doi.org/10.1002/ece3.3018

Duhem C, Vidal E, Legrand J, Tatoni T (2003) Opportunistic feeding responses of the Yellow-legged Gull Larus michahellis to accessibility of refuse dumps: the gulls adjust their diet composition and diversity according to refuse dump accessibility. Bird Study 50:61–67. https://doi.org/10.1080/00063650309461291

Duhem C, Vidal E, Roche P, Legrand J (2005) How is the diet of yellow-legged gull chicks influenced by parents' accessibility to landfills? Waterbirds 28:46–52. https://doi.org/10.1675/1524-4695(2005)028[0046:HITDOY]2.0.CO;2

Elliott KH, Duffe J, Lee SL et al (2006) Foraging ecology of Bald Eagles at an urban landfill. Wilson J Ornithol 118:380–390. https://doi.org/10.1676/04-126.1

Espín S, Martínez-López E, María-Mojica P, García-Fernández AJ (2012) Razorbill (Alca torda) feathers as an alternative tool for evaluating exposure to organochlorine pesticides. Ecotoxicology 21:183–190. https://doi.org/10.1007/s10646-011-0777-z

Eulaers I, Covaci A, Herzke D et al (2011a) A first evaluation of the usefulness of feathers of nestling predatory birds for non-destructive biomonitoring of persistent organic pollutants. Environ Int 37:622–630. https://doi.org/10.1016/j.envint.2010.12.007

Eulaers I, Covaci A, Hofman J et al (2011b) A comparison of non-destructive sampling strategies to assess the exposure of white-tailed eagle nestlings (Haliaeetus albicilla) to persistent organic pollutants. Sci Total Environ 410–411:258–265. https://doi.org/10.1016/j.scitotenv.2011.09.070

Eulaers I, Jaspers VLB, Pinxten R et al (2014) Legacy and current-use brominated flame retardants in the Barn Owl. Sci Total Environ 472:454–462. https://doi.org/10.1016/j.scitotenv.2013.11.054

FAO/UNEP (1991) Informed decision guidance on DDT. Rome

Fazari WAAL, Mcgrady MJ (2016) Counts of Egyptian Vultures Neophron percnopterus and other avian scavengers at Muscat' s municipal landfill, Oman, November 2013–March 2015. 38:99–105

Fernie KJ, Shutt JL, Mayne G et al (2005) Exposure to polybrominated diphenyl ethers (PBDEs): changes in thyroid, vitamin A, glutathione homeostasis, and oxidative stress in American kestrels (Falco sparverius). Toxicol Sci 88:375–383. https://doi.org/10.1093/toxsci/kfi295

Fisk AT, Rosenberg B, Cymbalisty CD et al (1999) Octanol/water pertiotion coefficients of toaphene congeners determined by the "Slow-Stirring" method. Chemosphere 39:2549–2562

François A, Técher R, Houde M et al (2016) Relationships between polybrominated diphenyl ethers and transcription and activity of type 1 deiodinase in a gull highly exposed to flame retardants. Environ Toxicol Chem 35:2215–2222. https://doi.org/10.1002/etc.3372

Furness RW (1993) Birds as monitors of pollutants. In: Furness RW, Greenwood JJD (eds) Birds as monitors of environmental change. Springer, Dordrecht, The Netherlands, pp 86–143

Furness R (1997) Seabirds as monitors of the marine environment. ICES J Mar Sci 54:726–737. https://doi.org/10.1006/jmsc.1997.0243

García-Fernández AJ, Espín S, Martínez-López E (2013) Feathers as a biomonitoring tool of polyhalogenated compounds: a review. Environ Sci Technol 47:3028–3043. https://doi.org/10.1021/es302758x

Garthe S, Hüppop O (2004) Scaling possible adverse effects of marine wind farms on seabirds: developing and applying a vulnerability index. J Appl Ecol 41:724–734. https://doi.org/10.1111/j.0021-8901.2004.00918.x

Gentes ML, Mazerolle MJ, Giroux JF et al (2015) Tracking the sources of polybrominated diphenyl ethers in birds: foraging in waste management facilities results in higher DecaBDE exposure in males. Environ Res 138:361–371. https://doi.org/10.1016/j.envres.2015.02.036

Gilchrist TT, Letcher RJ, Thomas P, Fernie KJ (2014) Polybrominated diphenyl ethers and multiple stressors influence the reproduction of free-ranging tree swallows (Tachycineta bicolor) nesting at wastewater treatment plants. Sci Total Environ 472:63–71. https://doi.org/10.1016/j.scitotenv.2013.10.090

Goede AA, De Voogt P (1985) Lead and cadmium in waders from the Dutch Wadden Sea. Environ Pollut Ser A-Ecological Biol 37:311–322. https://doi.org/10.1016/0143-1471(85)90120-5

Gould JC, Cooper KR, Scanes CG (1999) Effects of polychlorinated biphenyls on thyroid hormones and liver type I monodeiodinase in the chick embryo. Ecotoxicol Environ Saf 43:195–203. https://doi.org/10.1006/eesa.1999.1776

Goutner V, Albanis T, Konstantinou I, Papakonstantinou K (2001) PCBs and organochlorine pesticide residues in eggs of Audouin's Gull (Larus audouinii) in the north-eastern Mediterranean. Mar Pollut Bull 42:377–388. https://doi.org/10.1016/S0025-326X(00)00165-X

Halbrook RS, Arenal CA (2003) Field studies using European starlings to establish causality between PCB exposure and reproductive effects. Hum Ecol Risk Assess 9:121–136. https://doi.org/10.1080/713609855

Hansen LG, Green D, Cochran J et al (1997) Chlorobiphenyl (PCB) composition of extracts of subsurface soil, superficial dust and air from a contaminated landfill. Fresenius J Anal Chem 357:442–448. https://doi.org/10.1007/s002160050186

Hario M, Rintala J (2016) Population trends in herring gulls (Larus argentatus), great black-backed gulls (Larus marinus) and lesser black-backed gulls (Larus fuscus fuscus) in Finland. Waterbirds 39:10–14. https://doi.org/10.1675/063.039.sp107

Harris MP (1971) Ecological adaptations of moult in some British gulls. Bird Study 18:113–118. https://doi.org/10.1080/00063657109476302

Hawker DW, Connell DW (1988) Octanol-water partition coefficients of polychlorinated biphenyl congeners. Environ Sci Technol 22:382–387. https://doi.org/10.1021/es00169a004

Hobson KA (1993) Trophic relationships among high Arctic seabirds: insights from tissue-dependent stable-isotope models. Mar Ecol Prog Ser 95:7–18. https://doi.org/10.3354/meps095007

Hobson KA, Clark RG (1992) Assessing avian diets using stable isotopes II: factors influencing diet-tissue fractionation. Condor 94:189–197

Hüppop O, Dierschke J, Exo K et al (2006) Bird migration studies and potential collision risk with offshore wind turbines. Ibis (Lond 1859). https://doi.org/10.1111/j.1474-919X.2006.00536.x

Huwe JK, Larsen GL (2005) Polychlorinated dioxins, furans, and biphenyls, and polybrominated diphenyl ethers in a U.S. meat market basket and estimates of dietary intake. Environ Sci Technol 39:5606–5611. https://doi.org/10.1021/es050638g

Ito A, Yamashita R, Takada H et al (2013) Contaminants in tracked seabirds showing regional patterns of marine pollution. Environ Sci Technol 47:7862–7867. https://doi.org/10.1021/es4014773

Jamieson AJ, Malkocs T, Piertney SB et al (2017) Bioaccumulation of persistent organic pollutants in the deepest ocean fauna. Nat Ecol Evol 1:51. https://doi.org/10.1038/s41559-016-0051

Jaspers VLB, Voorspoels S, Covaci A, Eens M (2006) Can predatory bird feathers be used as a non-destructive biomonitoring tool of organic pollutants? Biol Lett 2:283–285. https://doi.org/10.1098/rsbl.2006.0450

Jaspers VLB, Covaci A, Van den Steen E, Eens M (2007a) Is external contamination with organic pollutants important for concentrations measured in bird feathers? Environ Int 33:766–772. https://doi.org/10.1016/j.envint.2007.02.013

Jaspers VLB, Voorspoels S, Covaci A et al (2007b) Evaluation of the usefulness of bird feathers as a non-destructive biomonitoring tool for organic pollutants: a comparative and meta-analytical approach. Environ Int 33:328–337. https://doi.org/10.1016/j.envint.2006.11.011

Jaspers VLB, Covaci A, Deleu P et al (2008) Preen oil as the main source of external contamination with organic pollutants onto feathers of the common magpie (Pica pica). Environ Int 34:741–748. https://doi.org/10.1016/j.envint.2007.12.002

Jaspers VLB, Covaci A, Deleu P, Eens M (2009) Concentrations in bird feathers reflect regional contamination with organic pollutants. Sci Total Environ 407:1447–1451. https://doi.org/10.1016/j.scitotenv.2008.10.030

Jaspers VLB, Rodriguez FS, Boertmann D et al (2011) Body feathers as a potential new biomonitoring tool in raptors: a study on organohalogenated contaminants in different feather types and preen oil of West Greenland white-tailed eagles (Haliaeetus albicilla). Environ Int 37:1349–1356. https://doi.org/10.1016/j.envint.2011.06.004

Jaspers V, Megson D, O'Sullivan G (2014) POPs in the terrestrial environment. In: Environmental forensics for persistent organic pollutants. Elsevier B.V., pp 291–356

Johnson MS, Leah RT, Connor L et al (1996) Polychlorinated biphenyls in small mammals from contaminated landfill sites. Environ Pollut 92:185–191. https://doi.org/10.1016/0269-7491(95)00096-8

Kenyon Ross R (1976) Notes on the behavior of captive great cormorants. Wilson Bull 88:143–145

Kocagöz R, Onmus O, Onat I et al (2014) Environmental and biological monitoring of persistent organic pollutants in waterbirds by non-invasive versus invasive sampling. Toxicol Lett 230:208–217. https://doi.org/10.1016/j.toxlet.2014.01.044

Kruszyk R, Ciach M (2010) White Storks, Ciconia ciconia, forage on rubbish dumps in Poland—a novel behaviour in population. Eur J Wildl Res 56:83–87. https://doi.org/10.1007/s10344-009-0313-0

Laurin M, Gauthier JA (2012) Amniota. Mammals, reptiles (turtles, lizards, Sphenodon, crocodiles, birds) and their extinct relatives. http://tolweb.org/Amniota/14990/2012.01.30. Accessed 16 Aug 2017

Letcher RJ, Marteinson SC, Fernie KJ (2014) Dietary exposure of American kestrels (Falco sparverius) to decabromodiphenyl ether (BDE-209) flame retardant: Uptake, distribution, debromination and cytochrome P450 enzyme induction. Environ Int 63:182–190. https://doi.org/10.1016/j.envint.2013.11.010

Li B, Danon-Schaffer MN, Li LY et al (2012) Occurrence of PFCs and PBDEs in landfill leachates from across Canada. Water Air Soil Pollut 223:3365–3372. https://doi.org/10.1007/s11270-012-1115-7

Li Y, Li J, Deng C (2014) Occurrence, characteristics and leakage of polybrominated diphenyl ethers in leachate from municipal solid waste landfills in China. Environ Pollut 184:94–100. https://doi.org/10.1016/j.envpol.2013.08.027

Luo XJ, Zhang XL, Liu J et al (2009) Persistent halogenated compounds in waterbirds from an e-waste recycling region in south China. Environ Sci Technol 43:306–311. https://doi.org/10.1021/es8018644

MacKay D, Fraser A (2000) Bioaccumulation of persistent organic chemicals: mechanisms and models. Environ Pollut 110:375–391. https://doi.org/10.1016/S0269-7491(00)00162-7

Martínez-López E, Espín S, Barbar F et al (2015) Contaminants in the southern tip of South America: analysis of organochlorine compounds in feathers of avian scavengers from Argentinean Patagonia. Ecotoxicol Environ Saf 115:83–92. https://doi.org/10.1016/j.ecoenv.2015.02.011

Massemin-Challet S, Gendner JP, Samtmann S et al (2006) The effect of migration strategy and food availability on White Stork Ciconia ciconia breeding success. Ibis 148:503–508. https://doi.org/10.1111/j.1474-919X.2006.00550.x

Matache ML, Hura C, David IG (2016) Non-invasive monitoring of organohalogen compounds in eggshells and feathers of birds from the Lower Prut Floodplain Natural Park in Romania. Procedia Environ Sci 32:49–58. https://doi.org/10.1016/j.proenv.2016.03.011

McFarland VA, Clarke JU (1989) Environmental occurrence, abundance, and potential toxicity of polychlorinated biphenyl congeners: considerations for a congener-specific analysis. Environ Health Perspect 81:225–239. https://doi.org/10.1289/ehp.8981225

Melnyk A, Dettlaff A, Kuklińska K et al (2015) Concentration and sources of polycyclic aromatic hydrocarbons (PAHs) and polychlorinated biphenyls (PCBs) in surface soil near a municipal solid waste (MSW) landfill. Sci Total Environ 530–531:18–27. https://doi.org/10.1016/j.scitotenv.2015.05.092

Meyer J, Jaspers VLB, Eens M, de Coen W (2009) The relationship between perfluorinated chemical levels in the feathers and livers of birds from different trophic levels. Sci Total Environ 407:5894–5900. https://doi.org/10.1016/j.scitotenv.2009.07.032

Moreno R, Jover L, Munilla I et al (2010) A three-isotope approach to disentangling the diet of a generalist consumer: the yellow-legged gull in northwest Spain. Mar Biol 157:545–553. https://doi.org/10.1007/s00227-009-1340-9

Öman CB, Junestedt C (2008) Chemical characterization of landfill leachates—400 parameters and compounds. Waste Manag 28:1876–1891. https://doi.org/10.1016/j.wasman.2007.06.018

Patenaude-Monette M, Bélisle M, Giroux JF (2014) Balancing energy budget in a central-place forager: which habitat to select in a heterogeneous environment? PLoS One 9:1–12. https://doi.org/10.1371/journal.pone.0102162

Persson NJ, Pettersen H, Ishaq R et al (2005) Polychlorinated biphenyls in polysulfide sealants—occurrence and emission from a landfill station. Environ Pollut 138:18–27. https://doi.org/10.1016/j.envpol.2005.02.021

Poprach K, Machar I, Maton K (2016) Long-term decline in breeding abundance of Black-headed Gull (Chroicocephalus ridibundus) in the Czech Republic: a case study of a population trend at the Chomoutov lake. Ekológia (Bratislava) 35:350–358. https://doi.org/10.1515/eko-2016-0028

Qiu X, Zhu T, Yao B, et al (2005) Contribution of dicofol to the current DDT pollution in China. Environ Sci Technol 39:4385–4390. https://doi.org/10.1021/es050342a

Rajaei F, Sari AE, Bahramifar N et al (2011) Persistent organic pollutants in muscle and feather of ten avian species from māzandarān province of Iran, on the coast of the Caspian sea. Bull Environ Contam Toxicol 87:678–683. https://doi.org/10.1007/s00128-011-0420-y

Ruus A, Ugland KI, Skaare JU (2002) Influence of trophic position on organochlorine concentrations and compositional patterns in a marine food web. Environ Toxicol Chem 21:2356–2364. https://doi.org/10.1897/1551-5028(2002)021<2356:iotpoo>2.0.co;2

Schecter A, Colacino J, Patel K et al (2010) Polybrominated diphenyl ether levels in foodstuffs collected from three locations from the United States. Toxicol Appl Pharmacol 243:217–224. https://doi.org/10.1016/j.taap.2009.10.004

Scott P, Duncan P, Green JA (2014) Food preference of the Black-headed Gull Chroicocephalus ridibundus differs along a rural–urban gradient. Bird Study 1–8. doi: https://doi.org/10.1080/00063657.2014.984655

Smith PN, Johnson KA, Anderson TA, McMurry ST (2003) Environmental exposure to polychlorinated biphenyls among raccoons (Procyon lotor) at the paducah gaseous diffusion plant, Western Kentucky, USA. Environ Toxicol Chem 22:406–416. https://doi.org/10.1897/1551-5028(2003)022<0406:EETPBA>2.0.CO;2

Sol D, Arcos JM, Senar JC (1995) The influence of refuse tips on the winter distribution of Yellow-legged Gulls Larus cachinnans. Bird Study 42:216–221. https://doi.org/10.1080/00063659509477170

Soldatini C, Albores-Barajas YV, Torricelli P, Mainardi D (2008) Testing the efficacy of deterring systems in two gull species. Appl Anim Behav Sci 110:330–340. https://doi.org/10.1016/j.applanim.2007.05.005

Sommerfeld J, Mendel B, Fock HO, Garthe S (2016) Combining bird-borne tracking and vessel monitoring system data to assess discard use by a scavenging marine predator, the lesser black-backed gull Larus fuscus. Mar Biol 163:1–11. https://doi.org/10.1007/s00227-016-2889-8

Stockholm Convention (2009) Stockholm convention on Persistent Organic Pollutants (POPs). In: Fourth meeting of the Conference of the Parties (Decisions SC-4/10 to SC-4/18). Geneva, Switzerland, pp 1–64

Stockholm Convention (2011) An amendment to annex A adopted by the conference of the parties to the Stockholm convention on persistent organic pollutants at its fifth meeting (Decision SC-5/3). 2

Stockholm Convention (2013) An amendment to annex A adopted by the conference of the parties to the Stockholm convention on persistent organic pollutants at its fi sixth meeting (Decision SC-5/3). 15:2

Stockholm Convention (2015) Amendments to annexes A and C adopted by the conference of the parties to the Stockholm convention on persistent organic pollutants at its seventh meeting (Decisions SC—7/12, SC—7/13 and SC—7/14). 2–4

Strandberg B, Bandh C, Van Bavel B et al (1998) Concentrations, biomagnification and spatial variation of organochlorine compounds in a pelagic food web in the northern part of the Baltic Sea. Sci Total Environ 217:143–154. https://doi.org/10.1016/S0048-9697(98)00173-9

Summers JW, Gaines KF, Garvin N et al (2010) Feathers as bioindicators of PCB exposure in clapper rails. Ecotoxicology 19:1003–1011. https://doi.org/10.1007/s10646-010-0481-4

Tauler-Ametller H, Hernández-Matías A, Pretus JL, Real J (2017) Landfills determine the distribution of an expanding breeding population of the endangered Egyptian Vulture Neophron percnopterus. Ibis. https://doi.org/10.1111/ibi.12495

Técher R, Houde M, Verreault J (2016) Associations between organohalogen concentrations and transcription of thyroid-related genes in a highly contaminated gull population. Sci Total Environ 545–546:289–298. https://doi.org/10.1016/j.scitotenv.2015.12.110

Tyson C, Shamoun-Baranes J, Van Loon EE et al (2015) Individual specialization on fishery discards by lesser black-backed gulls (Larus fuscus). ICES J Mar Sci 69:84–88. https://doi.org/10.1093/icesjms/fsr174

Van den Steen E, Covaci A, Jaspers VLB et al (2007) Experimental evaluation of the usefulness of feathers as a non-destructive biomonitor for polychlorinated biphenyls (PCBs) using silastic implants as a novel method of exposure. Environ Int 33:257–264. https://doi.org/10.1016/j.envint.2006.09.018

Van den Steen E, Jaspers VLB, Covaci A et al (2008) Variation, levels and profiles of organochlorines and brominated flame retardants in great tit (Parus major) eggs from different types of sampling locations in Flanders (Belgium). Environ Int 34:155–161. https://doi.org/10.1016/j.envint.2007.07.014

Voorspoels S, Covaci A, Jaspers VLB et al (2007) Biomagnification of PBDEs in three small terrestrial food chains. Environ Sci Technol 41:411–416. https://doi.org/10.1021/es061408k

Watanabe MX, Iwata H, Watanabe M et al (2005) Bioaccumulation of organochlorines in crows from an Indian open waste dumping site: evidence for direct transfer of dioxin-like congeners from the contaminated soil. Environ Sci Technol 39:4421–4430. https://doi.org/10.1021/es050057r

Wetmore A (1920) The Function of Powder Downs in Herons. Condor 22:168–170

Yamashita R, Takada H, Murakami M et al (2007) Evaluation of noninvasive approach for monitoring PCB pollution of seabirds using preen gland oil. Environ Sci Technol 41:4901–4906

Zeng YH, Luo XJ, Tang B, Mai BX (2016) Habitat- and species-dependent accumulation of organohalogen pollutants in home-produced eggs from an electronic waste recycling site in South China: levels, profiles, and human dietary exposure. Environ Pollut 216:64–70. https://doi.org/10.1016/j.envpol.2016.05.039

Zhao Z, Jia J, Wang J et al (2018) Pollution levels of DDTs and their spatiotemporal trend from sediment records in the Southern Yellow Sea, China. Mar Pollut Bull 127:359–364. https://doi.org/10.1016/j.marpolbul.2017.12.026

Zupancic-Kralj L, Jan J, Marsel J (1992) Assessment of polychlorobiphenyls in human/poultry fat and in hair/plumage from a contaminated area. Chemosphere 25:1861–1867

The Environmental Behavior of Methylene-4,4′-dianiline

Thomas Schupp, Hans Allmendinger, Christian Boegi, Bart T. A. Bossuyt, Bjoern Hidding, Summer Shen, Bernard Tury, and Robert J. West

Contents

T. Schupp (✉)
Faculty of Chemical Engineering, Muenster University of Applied Science, Steinfurt, Germany
e-mail: thomas.schupp@fh-muenster.de

H. Allmendinger
Currenta GmbH & Co. OHG, Leverkusen, Germany
e-mail: hans.allmendinger@currenta.de

C. Boegi
BASF SE, FEP/PA - Z570, Ludwigshafen, Germany
e-mail: christian.boegi@basf.com

B. T. A. Bossuyt
Huntsman Europe, Everberg, Belgium
e-mail: bart_bossuyt@huntsman.com

B. Hidding
BASF SE, RB/TC - Z570, Ludwigshafen, Germany
e-mail: bjoern.hidding@basf.com

S. Shen
Dow Chemical (China) Investment Limited Company, Shanghai, China
e-mail: SMShen2@dow.com

B. Tury
(former) International Isocyanate Institute Inc., Boonton, NJ, USA
e-mail: berntury@yahoo.com

R. J. West
International Isocyanate Institute Inc., Boonton, NJ, USA
e-mail: bob@iiiglobal.net

© Springer International Publishing AG, part of Springer Nature 2018   91
P. de Voogt (ed.), *Reviews of Environmental Contamination and Toxicology*
Volume 246, Reviews of Environmental Contamination and Toxicology 246,
DOI 10.1007/398_2018_13

1   Introduction

4,4′-Methylenedianiline (MDA; CAS-No. 101-77-9) is a high production volume chemical with annual world production volume estimated to exceed four million metric tons per year (Carvajal-Diaz 2015). More than 98% of this production is consumed as an intermediate for the production of the methylenediphenyl diisocyanate (MDI) substances, which are important monomers for the versatile thermoset polymer group of polyurethanes. Minor amounts of the MDA are also consumed in manufacture of high performance polyimide fibers, and in manufacture of other specialty chemicals and resins.

The industrial-scale production of the MDA occurs *via* a condensation reaction of aniline and formaldehyde, as shown in Fig. 1. The bulk product of this reaction is commonly referred to as polymeric methylenedianiline (pMDA), and the relative proportions of the illustrated components can be controlled by adjusting the ratio of aniline and formaldehyde reactants. It should be noted here that the name "polymeric MDA" does not necessarily indicate that this reaction mixture meets the current OECD definition of "polymer". In many instances, due to >50% of composition coming from 4,4′-methylenedianiline, the "polymeric MDA" reaction product would not meet this OECD definition (http://www.oecd.org/env/ehs/oecddefinitionofpolymer.htm). The pMDA reaction products can be further isolated or purified by fractional distillation, making possible any number and combination of the substances listed in Table 1 (Alport et al. 2003). The 2:1 condensation products are sometimes called "2-ring-MDA" but are hereafter referred to as methylenedianiline (MDA). When isolated from the bulk pMDA reaction mixture, the 2-ring MDA will typically occur as a mixture of three positional isomers (Fig. 2), where the 4,4′-MDA is the predominant isomer representing more than 90–95% of the isomer mixture, with the 2,4′-MDA and

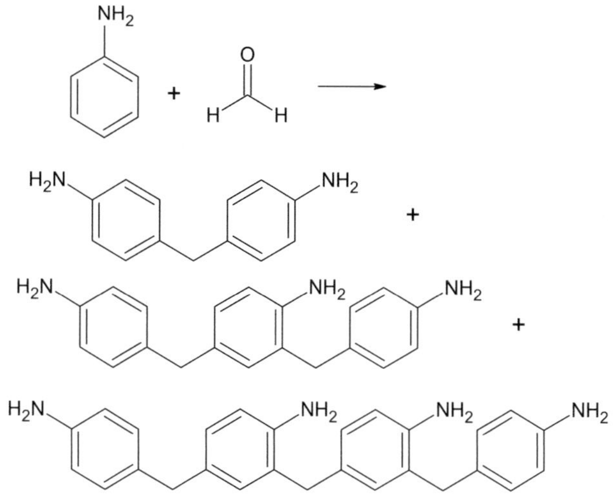

Fig. 1 Generalized commercial synthesis route of the methylenedianiline substances (MDAS)

Table 1 Identity of the commercially relevant methylenedianiline substances (MDAS)

Substance	Chemical name	Chemical abstracts registry number	Synonyms	Typical % in MDA	Typical % in pMDA
Polymeric Methylene-dianiline (pMDA)	Formaldehyde, polymer with benzenamine	25214-70-4	pMDA	N/A	100
Methylenedianilines (MDA)	Benzenamine, 4,4'-methylenebis	101-77-9	4,4'-MDA	90–95	50
	Benzeneamine, 2, 4'-methylenebis	1208-52-2	2,4'-MDA	2–5	
	Benzeneamine, 2,2'-methylenebis	6582-52-1	2,2'-MDA	<1	
Oligomeric Methylenedianilines	3-ring	N/A	oMDA	N/A	25
	4-ring			N/A	12
	5-ring			N/A	6
	Higher oligomers			N/A	<6

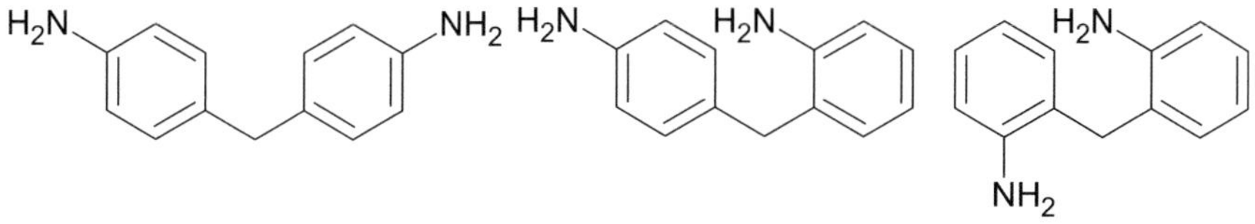

Fig. 2 Molecular structures of the 4,4'-, 2,4'-, and 2,2'-isomers of methylenedianiline

2,2'-MDA making up 2–5% and less than 1% of the mixture, respectively. The higher oligomer components of pMDA are sometimes named "3-ring-MDA," "4-ring-MDA," and so on and are hereafter referred to as oligomeric MDA or oMDA.

In the evaluation of physical–chemical and toxicological properties of chemical substances, it is desirable to conduct testing on commercially relevant substances which occur at a high purity or as a single component. For this reason, and because it is among the most commercially prominent of the methylenedianiline substances, the 4,4'-MDA substance has been intensely investigated for its physical–chemical and toxicological properties. Concerning environmental behavior, it was almost always the 4,4'-MDA that was used as test substance. In 2001, the European Union issued the EU risk assessment report on 4,4'-MDA which provided a summary of all physical–chemical and hazard property data known by then (European Union 2001). Since that time, and in preparation for the registration under commission regulation 1907/2006 in the EU, some more data on 4,4'-MDA were generated. Robust summaries of these past and more recent studies are now available to the public *via* the European Chemicals Agency (ECHA) web site (http://echa.europa.eu/web/guest/information-on-chemicals/registered-substances). In addition, the government of Canada has recently completed a draft screening level risk assessment for a grouping of the MDA, pMDA, and MDI substances (Government of Canada 2014). These and other regulatory assessments can provide good overviews of the available physical–chemical, health, and environmental properties of the MDAS family, as well as their assessed potential exposures and risks. This current review provides a deeper focus and summary of the known environmental properties of MDA as the most abundant and important member of the MDAS family as available from published and private company studies available through year 2014. A similar in-depth review and summary of the ecotoxicological properties for the MDAS family are reviewed in a previous review article of this journal (Schupp et al. 2016).

The majority of studies cited in this review and in the aforementioned regulatory assessments of the MDAS have been commissioned by the International Isocyanates Institute, Inc. (III; http://diisocyanates.org/) and its private company members and have not been published previously. For such cited references which are not from publically available sources, the ECHA web site (ECHA 2014a, b) provides robust study summaries for these studies.

2 Physical–Chemical Properties

The physical–chemical properties of 4,4'-MDA and pMDA are summarized in Table 2. These data for 4,4'-MDA are summarized in the EU risk assessment report (European Union 2001) and in the more recent robust study summaries provided by ECHA (2014a). The substance tested consisted of 97.39% 4,4'-MDA, 1.98% 2,4'-MDA, 50 ppm 2,2'-MDA, and less than 10 ppm aniline.

Physical–chemical properties for pMDA are not that uniform and depend on the composition, as pMDA is an oligomeric mixture which varies slightly in composition

Table 2 Physical–chemical properties of the 4,4′-methylenedianiline (MDA) and polymeric methylenedianiline (pMDA) substances

Property	4,4′-Methylenedianiline (MDA)[a]	Polymeric Methylenedianiline (pMDA)[b, c]
Physical state at 20°C	White solid	Viscous liquid
Melting point (°C)	90–92	Glass transition at 0.8 and −2.7
Boiling point (°C)	398 ± 5 @ 101.3 kPa	410.6 @ 101.3 kPa
Bulk density (g/cm³, 20°C)	1.150	1.150
Vapor pressure (Pascal)	2.5×10^{-4} (25°C)	$<1 \times 10^{-4}$ (20°C) 1.6×10^{-4} (50°C)
Water solubility (g/L at 25°C)	pH 5.3 = 2.2; pH 7 = 1.01; pH 9 = 0.84	pH 7 = 0.36–1.22
Octanol–water partition coefficient (log K_{ow})	1.55	1.2–2.7
Dissociation constant (pKa at 20°C)	4.96	Not determined

[a]ECHA (2014a)
[b]ECHA (2014b)
[c]pMDA sample containing 58% 2-ring-, 23% 3-ring-, 10% 4-ring-, and 3% 5-ring-isomers; 6% higher oligomers

based on the ratio of its reactants (Table 1). Robust study summaries of these pMDA properties are also provided by ECHA (2014b).

As weak bases, the water solubilities of the MDAS depend on the pH. Though the individual molecules possess two or more primary amino groups, their respective pKa values are essentially indistinguishable by the spectrophotometric method of OECD Guideline 112. The apparent equivalence of the pKa values for these amino groups is owed to the fact that delocalization of the positive charge of the corresponding ammonium ions across the molecule is restricted by the methylene group which bridges the aromatic rings. The measured value is in excellent agreement with the pKa values derived by the SPARC calculator (ARChem LLC, 2017; http://archemcalc.com/sparc-web/calc.):

$$(MDA)H_2^{2+} \rightleftarrows H^+ + (MDA)H^+; \quad pKa1 = 4.76;$$
$$(MDA)H^+ \rightleftarrows H^+ + MDA; \quad pKa2 = 4.97.$$

The abundance of MDA, $(MDA)H^+$ and $(MDA)H_2^{2+}$, in dependence of the pH is shown in Fig. 3.

Deng et al. (2015) published data on the solubility of 4,4′-MDA in several organic solvents at different temperatures. For example, at 298.15 K, molar fractions of 4,4′-MDA at saturation are 12.93% in methanol, 4.61% in ethanol, 1.36% in 2-propanol, 2.35% in 1-butanol, 2.99% in benzene, 1.95% in toluene, and 14.23% in chloroform.

Fig. 3 Ratio of MDA (S1) to (MDA)H$^+$ (S2) and (MDA)H$_2$$^{2+}$ (S3) in dependence on the pH value

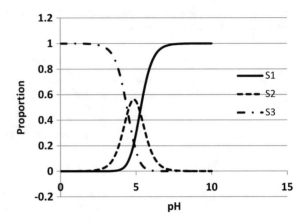

3 Biodegradation Screening Tests and Behavior in Water and Sediment

MDA was investigated for ready and inherent biodegradability in OECD screening tests and in further simulation biodegradation tests. Further, tests for abiotic transformation of MDA were performed. Data on degradation of MDA in screening tests and in tests simulating the aquatic environment are summarized in Table 3.

3.1 Tests for Ready Biodegradability

Baumann (1985) performed an OECD 301B test with MDA. When the initial load was 10 mg/L, the yield of CO_2 was 10% after 28 days, but only 2% when the initial load was 20 mg/L.

MDA was not degraded when tested according to OECD 301C (CITI 1992). The inoculum used was generated from sludge from ten different municipal sewage treatment plants (STPs), fed for at least 1 month with synthetic sewage.

In an OECD test guideline 301F test, in one of three replicates MDA was degraded by 19% according to BOD analysis, and 34% due to DOC removal (Yakabe 1994). In the other two replicates, there was no degradation of MDA. Municipal STP sludge was applied as inoculum.

Ekici et al. (2001a) followed primary disappearance of 4,4′-MDA by GC-MS analysis in batches with municipal sludge. Fresh activated sludge from a municipal STP was centrifuged and resuspended with tap water to yield a content of 1 g dry substance per liter. 25 mL of this suspension were incubated with 50 μg/L MDA for 7 days at 35°C, and CO_2 was trapped in Ba(OH)$_2$. For anaerobic tests, nitrogen was

Table 3 Results of screening biodegradation tests with MDA and simulation tests in aquatic environments

Guideline	Result	Remarks	Reference
OECD 301B	2%/10% CO_2	MDA: 20 mg/L and 10 mg/L	Baumann (1985)
OECD 301C	0% BOD		CITI (1992)
OECD 301F	19%/34% O_2	BOD/DOC	Yakabe (1994)
OECD 301B	46% CO_2	0.5 mg/L MDA, radiolabeled	Schwarz (2009)
OECD 301A	94.8% DOC		Mei et al. (2015)
OECD 301B	29.5% CO_2		Mei et al. (2015)
OECD 301D	0% BOD		Mei et al. (2015)
OECD 301F	100% O_2		Mei et al. (2015)
OECD 302C	43% BOD		Caspers et al. (1986)
OECD 302B	>95% DOC	Abstract only	BASF (1981)
OECD 303A	6% DOC		Baumann (1986)
OECD 309	18% CO_2; $DT_{50} = 11$ days	Estuarine water, 92 days	Schaefer and Carpenter (2013)
OECD 309	25.5% CO_2; $DT_{50} = 7.5$ days	Freshwater, 92 days	Schaefer and Carpenter (2013)
OECD 308	2.8% CO_2; $DT_{50} = 5.8$ days	Estuarine water, aerobic, 100 days	Schaefer and Ponizovsky (2013)
OECD 308	5.7% CO_2; $DT_{50} = 3.1$ days	Freshwater sediment, aerobic, 100 days	Schaefer and Ponizovsky (2013)
OECD 308	0.64% CO_2; 0.032% VOC; $DT_{50} = 21$ days	Estuarine water sediment, anaerobic, 102 days	Schaefer and Ponizovsky (2013)
OECD 308	0.58% CO_2; 0.09% VOC; $DT_{50} = 10$ days	Freshwater sediment, anaerobic, 102 days	Schaefer and Ponizovsky (2013)
-----	$k = 3.3 \times 10^{-2}\,h^{-1}$/ $k = 2.7 \times 10^{-9}\,h^{-1}$	Primary degradation aerobic conditions/under argon	Ekici et al. (2001a)
US EPA 2004	$k = 5.7 \times 10^{-3}\,h^{-1}$/ $k < 10^{-3}\,h^{-1}$	Primary degradation at 3% NaCl/7% NaCl	Suidan et al. (2011)

bubbled for 1 h through the sludge suspension before MDA was added. After certain time intervals, samples were taken, mixed with 1 N NaOH, and extracted with ethyl acetate; the extract was concentrated to dryness at 50°C/15 mmHg and the residue was dissolved in acetone. The acetone solution was taken for GC analysis of MDA. The half-life for 4,4′-MDA was 21.3 h under aerobic conditions, and less than 30 min in the same system when purged with nitrogen. MDA metabolites were not identified. The authors provide no explanation for the difference in half-lives between aerobic and anaerobic conditions. Although Ba(OH)$_2$ traps were used for CO_2, CO_2 yields were not reported.

Schwarz (2009) performed an OECD 301B degradation study and made use of radiolabeled MDA with a loading of 0.5 mg/L. CO_2 production increased after an adaptation time of about 10 days. The yield of labeled CO_2 was 46% after 28 days, and 53% after 63 days. At the end of the test, 3.7% of the test substance was dissolved in the aquatic phase, and about 14% of the initial radioactivity was absorbed on sludge. 0.5 mg/L MDA did not inhibit the degradation of the positive control aniline under the same test conditions.

Mei et al. (2015) ran OECD 301A, 301B, 301D, and 301F tests with MDA. The initial MDA concentration for the OECD 301A, 301B, and 301F tests ranged from 15.8 to 40 mg/L while an inoculum of about 30 mg/L (dry wt.) was used. The OECD 301D test was carried out with filtrate from activated sludge with about 10^6 colony forming units (CFU)/L and an initial MDA concentration of 2 mg/L. The inoculum originated from the Guangzhou wastewater treatment plant, China. From the data provided, it is not clear whether or not the wastewater treatment plant has a history of MDA exposure. Blank and positive controls (sodium benzoate) as well as toxicity controls (MDA plus sodium benzoate) were run in parallel. For the OECD 301A test, a sterile control where microbial activity was inhibited by 10 mg/L $HgCl_2$ was performed in addition. MDA did not inhibit the biodegradation of benzoate. In the OECD 301D and 301F tests, there was no increase in nitrite and nitrate, so nitrification was not significant. In the OECD 301D test, neither oxygen consumption nor primary decay of MDA was observable. MDA decay was >99% in the remaining tests; lag times were between 11 and 13 days (which does not indicate a history of MDA exposure), and the degradation reached 95% (DOC removal, OECD 301A), 30% (CO_2 evolution, OECD 301B), and 100% (O_2 consumption, OECD 301F) on day 28. Concerning the results of the 301A and 301F test, MDA would be regarded as readily biodegradable, whereas results from the OECD 301B and 301D test show no ready biodegradability. The authors argue that the results from the OECD 301D tests are of limited value as due to filtration of sludge and the very low load of organic material, certain microorganisms were probably excluded and those present were exposed to rather unfavorable conditions. In the 301A, 301B, and 301D tests, an intermediate yellow discoloration was observable which remained in the 301B test. The authors hypothesize that the difference between the 301A and 301F test on the one hand and the 301B test on the other hand is attributable to the usage of the MDA molecule for the buildup of biomolecules; this would explain the disappearance of DOC and consumption of oxygen while the formation of carbon dioxide is poor.

3.2   Test for Inherent Biodegradability

Caspers et al. (1986) report 43% degradation (BOD) of MDA when tested according to OECD guideline 302C. The inoculum used originated from a laboratory plant, which was run predominantly with municipal sewage.

In an abstract, MDA was said to be completely eliminated (>95%) within 27 days when exposed to an inoculum from an adapted, industrial STP according to OECD 302B (BASF 1981). After 3 h, the DOC removal achieved 8%.

In a STP simulation test according to OECD guideline 303A, the mean removal of MDA was 6.5%, only (Baumann 1986).

Suidan et al. (2011) performed a degradation test with MDA in hypersaline water according to the US Environmental Protection Agency guidelines for assessing biodegradability in biological treatment units, US EPA 2004. Industrial effluents containing 3 or 7% NaCl were loaded with 5 mg/L MDA, and the decay of MDA was analyzed by HPLC-MS. At 3% NaCl, the first-order rate constant for MDA removal was 0.0057 h^{-1}; at 7%, there was no noticeable removal of MDA over the testing period of 120 h. The disappearance of sodium acetate, however, was not influenced by MDA.

3.3 Simulation Tests and Ultimate Mineralization in Environmental Samples

Schaefer and Carpenter (2013) performed a degradation test in surface water samples according to OECD Guideline No. 309. Radiolabeled MDA was added to a concentration of 100 µg/L. In the estuarine water sample, after 92 days 18% CO_2 was generated, MDA was removed by more than 99%, and concerning the disappearance of MDA 50% was removed after 11 days and 90% after 37 days. In the freshwater sample, 25.5% CO_2 were formed after 92 days, MDA was removed by more than 99%, and primary removal was 50% after 7.5 days and 90% after 25 days. In killed controls, only about 0.2% CO_2 was generated, and the removal of MDA was 22.6% or 42.8% in the estuarine and freshwater sample, respectively. The higher yield of CO_2 in the freshwater sample against the estuarine water goes hand in hand with more than sevenfold higher numbers of bacterial and fungal CFU. Water was extracted with ethyl acetate for chemical analysis (HPLC-β-RAM) at several points in time. In the ethyl acetate extracts, besides MDA there were five other peaks more polar than MDA, and three less polar than MDA. In the extracted water, three metabolites more polar than MDA were detectable. In the sterile controls, two metabolites more polar than MDA were detectable as well. This investigation demonstrates that MDA can be mineralized in the presence of bacteria in environmental water samples, but several metabolites are formed as well, most of them attributable to microbial activity.

Schaefer and Ponizovsky (2013) looked into the degradability of MDA in environmental water and sediment systems. The estuarine water sample was associated with loamy sediment with more than 50% of clay and an organic carbon content of 5.7–6.6%, whereas the freshwater sample was combined with sandy sediment with an organic carbon content of 0.4–0.57%. The freshwater sample including sediment had more than sevenfold bacterial CFU than the estuarine samples. In the aerobic samples, MDA disappeared after 7 or 14 days, and 5.7 or 2.8% CO_2 were formed after 100 days (freshwater against estuarine water). Less than 0.05% was attributable to volatile compounds other than CO_2. For the estuarine and freshwater sediment systems, after 100 days 6.1 and 11.7% of the applied radioactivity remained as unidentified metabolites in water, 8.4 and 12% were extractable from sediment, and 90.2 and 69.3% were non-extractable from the sediment for the estuarine and freshwater systems, respectively. In the anaerobic systems, after 102 days only about 0.6% CO_2 and 0.05% VOC had been formed. In the estuarine system, 18.6% of the applied radioactivity was retrievable in the water, 32.9% were extractable from the sediment, and 85.2% were non-extractable in the sediment (sums up to more than 130%!). In the freshwater system, 12.1% of the applied radioactivity was retrievable in the water, 29.9% were extractable from the sediment, and 53.7% were non-extractable in the sediment. Under aerobic conditions, the time for 50% disappearance (DT_{50}) of MDA was 5.8 days for the estuarine, and 3.1 days for the freshwater system; interestingly, in sterile controls, DT_{50} was lower for the estuarine system (4.1 days) and higher for the freshwater system (3.8 days). Anaerobic conditions resulted in DT_{50} values of 21 days and 10 days for the estuarine and freshwater system, respectively; the decay of MDA did no longer follow first-order kinetics. In sterile controls under anaerobic conditions, DT_{50} was 5 days in both systems.

3.4   Abiotic Reactions in Water

Ekici et al. (2001a) reported half-life times for MDA hydrolysis of more than 12 years at 25°C. Hydrolysis, therefore, is not an important way for MDA disappearance in the environment.

Hellpointer (1997) investigated the decay of MDA in water under simulated solar irradiation. MDA has an absorption coefficient of 21,665 L/mol/cm at $\lambda = 243$ nm, and 3,497 L/mol/cm at $\lambda = 287$ nm. The quantum yield for MDA in polychromatic light is about 0.006 in the range 295–400 nm. According to the GC-solar program, the half-life time was calculated as 2.3 days and 3.8 days in summer for the 30th and 60th latitude, respectively. For winter, the corresponding half-life times are 7.1 days and 299 days, respectively. According to the solar irradiation in Central Europe as reported by Frank and Kloepffer (1988), the calculated half-life time for the 50th latitude is 15 days for July and >1 year for January. The tests were performed at pH = 5, 7, and 9; the pH of the water had no influence on the test results. Besides MDA, several peaks were detectable in the HPLC-UV runs at lower and higher retention times than MDA.

Ekici et al. (2001b) scrutinized the behavior of different aromatic amines in water–methanol mixtures (2:1) at 0.01 mol/L under UV light. They reported an extinction coefficient of 0.0065 L/mol/cm for MDA at $\lambda = 290$ nm. In degassed and argon saturated water, the quantum yield was zero, but 0.0049 when the water was saturated with oxygen after degassing (no more experimental details provided concerning degassing and oxygen saturation of the water). For MDA, the authors report a half-life time of 345 min under experimental conditions, and an environmental life time of 554 days. Parallel tests with trapping agents revealed that the reaction proceeds via excitation of the amine which consequently reacts with oxygen. Photo-oxidation product is 4,4'-diaminobenzophenone which then can form 4,4'-diaminobiphenyl by cheletropic scission. The latter reaction is enhanced by triplet state sensitizers like acetone or phenanthrene, but not by those with lower triplet energy, and it is nearly blocked in oxygen saturated water.

4 Behavior in Soil

In soil, MDA undergoes different reactions like adsorption, absorption, metabolism, and mineralization. These pathways are addressed in different subchapters.

4.1 Studies on Adsorption and Binding to Soil

Cowen et al. (1996) performed tests on adsorption and desorption of MDA from two soil samples. First, they checked different solvents for the usability for MDA extraction and stability of MDA in solvents. MDA was stable in different solvents like acetonitrile, tetrahydrofuran, acetone, dichloromethane, or aqueous ammonia solution for at least 28 days. Therefore, these solvents could be used for soil extraction experiments without generating any loss of MDA. For testing the extraction conditions, silt loam soil was placed in a glove box and flushed with nitrogen for 4 h to remove oxygen. Then, MDA was added with water to give 75% of 1/3 bar water tension and shaken for 18 h. After this time, 10 g dry Na_2SO_4 were added and the mixture was subjected to Soxhlet extraction with methanol or dichloromethane for 48 h; after this time, the solvent was evaporated to dryness and the soil was extracted with 50% NaOH under reflux for 2 h. As alternative to Soxhlet extraction, soil samples were mixed with solvent and sonicated for 20 h. Recovery of MDA after sonication was quantitative when MDA was contacted to sand which contained neither organic matter nor clay minerals; tetrahydrofuran and 10% methanol in 0.2 N H_2SO_4 were equally effective. In case of kaolinite, only 3% of MDA could be extracted with methanol in sulfuric acid after 20 h sonication. In case of silt loam, 97% acetonitrile with 3% of 30% ammonia solution was the best solvent for sonication extraction, but only 11% of MDA could be recovered. For Soxhlet extraction, methanol was better than dichloromethane, giving 17% MDA recovery.

Cowen et al. (1996, 1998) investigated the soil adsorption/desorption of radiolabeled MDA in the scope of terrestrial biodegradation experiments. Autoclaved sandy loam or silt loam was shaken with different concentrations of radiolabeled MDA in 0.01 M $CaCl_2$ at 23°C. A pretest has shown that a rapid adsorption phase is finished after about 4 h. After 8 h, 3 days, and 7 days, samples were centrifuged and radioactivity was measured in the supernatant. The Freundlich isotherm plot

$$\log(C_{\text{soil}}) = \log(K_d) + \frac{1}{n} \times \log(C_{\text{water}})$$

delivered the distribution coefficient, K_d, as intercept of the straight line with the ordinate. K_{oc} is calculated from K_d through division by organic matter content. The authors report that from 8 h to 7 days equilibration time, the apparent K_{oc} increases. For aerobic conditions, the log K_{oc} values for the two soils were 3.75 and 3.60 after 8 h, and 3.93 and 3.75 after 7 days. Under anaerobic conditions, 8-h log K_{oc} was 3.58 for both soils, and the 7-day log K_{oc}'s were 4.05 and 4.01. Anaerobic conditions seem to result in more efficient MDA absorption. Table 4 summarizes the data in dependence on soil type, time, and aerobic condition. The two soil types differ slightly in cation exchange capacity (CEC), total organic carbon (TOC) content, and pH. Although the pH difference between the two soils used is 0.8 units only, the influence on the abundance of the neutral, mono-protonated, and di-protonated MDA is strong. At pH $= 5.0$, the relation $MDA:MDAH^+:MDAH_2^{2+}$ is 0.292:0.55:0.158; at pH $= 5.8$, the relation is 0.762:0.227:0.01. That is, in the soil with the higher CEC, MDA is mostly present as non-charged molecule, and vice versa. The pH-dependent log K_{oc} can be estimated with the SPARC calculator (ARChem LLC, 2017; http://archemcalc.com/sparc-web/calc); for the soils used, with 90 mM bivalent cations and 6 mM monovalent cations, the calculated log K_{oc} is 5.86 for pH $= 5.0$ and 5.04 for pH $= 5.8$, two orders of magnitude higher than the measured values. This result casts doubt on the use of the calculated values for MDA for environmental modeling on a first glance; however, the calculated log K_{oc} of 3.737 at pH $= 7$ fits in the range of measured 8-h log K_{oc} values. Cowen et al. (1996) reported the soil pH values as part of the soil characterization; pH values of test mixtures were not reported. If it is assumed that in the test batches the pH was likely to approach 7, calculated and experimental 8-h log K_{oc} values are in excellent agreement.

Further decay of MDA in the aqueous solution between 8 h and 7 days – and subsequently increase in the apparent K_{oc} – was interpreted as irreversible absorption to soil organic matter. To check this assumption, in a separate experiment 7.5 mg/L MDA was shaken with Aldrich humic acid solution (11.4 g/L) for 1.5 h at 23°C;

Table 4 K_d and K_{oc} values for 4,4′-MDA (Cowen et al. 1996, 1998)

| | Sandy loam CEC[a] $= 177$; TOC[b] $= 1.6$; pH $= 5.8$ | | | | Silt loam CEC $= 136$; TOC $= 1.3$; pH $= 5$ | | | |
| | Aerobic | | Anaerobic | | Aerobic | | Anaerobic | |
Contact time	K_d	K_{oc}	K_d	K_{oc}	K_d	K_{oc}	K_d	K_{oc}
8 h	90.9	5,681	61.2	3,825	52.2	4,015	49.8	3,831
7 days	135	8,413	179	11,158	74	5,669	134	10,300

[a]Cation exchange capacity (CEC) (mM/kg)
[b]Total organic carbon content (%)

then, the humic acid was precipitated by bringing the solution to pH $= 0.8$ with HCl; in the filtrate, MDA was not detectable by HPLC. For toluene-2,6-diamine, more than 90% of the initially added substance was detectable in the filtrate under identical test conditions. That means that rather than covalent binding cation exchange is responsible for the MDA disappearance, and that the short contact time of 1.5 h was not sufficient for complete reaction of the sterically hindered primary amino groups of the toluene-2,6-diamine. In another separate experiment, radiolabeled toluene-2,6-diamine was equilibrated with Montmorillonite; 4 h contact were sufficient to achieve equilibrium. These results support the interpretation of a first "rapid" equilibrium partitioning, followed by a slower process which is thought to be covalent binding to soil organic matter. Reactions of primary aromatic amines with soil organic matter under formation of covalent C–N bonds are reported in the literature (p. e. Parris 1980; Saxena and Bartha 1983; Graveel et al. 1985; Ononyne and Graveel 1994).

In the desorption tests with MDA, Cowen et al. (1996) isolated the soil samples after they had 3 days or 7 days contact to MDA solution by centrifugation; then the soil pellets were equilibrated with fresh 0.01 M CaCl$_2$ solution for 24 h. For the 3 days soil samples, about 15% and 20% of the radioactivity could be desorbed from silt loam under aerobic and anaerobic conditions, respectively; for silt loam shaken with MDA solution for 7 days under aerobic and anaerobic conditions, desorption yielded 9% and 13% of the bound radioactivity. For sandy loam under aerobic conditions, desorption yields were about 15% and 13% for samples equilibrated previously with MDA for 3 days and 7 days under aerobic conditions; for anaerobic conditions, the corresponding yields for 3 days and 7 days samples were about 12% and 9%. Anaerobic conditions seem to favor irreversible adsorption of MDA to soil. For toluene-diamines, the authors report higher rates of irreversible absorption under aerobic against anaerobic conditions.

West et al. (2002) performed an extended anaerobic biodegradation study with radiolabeled MDA in either aquifer sand or loamy sand. 25 g soil (dry weight) were mixed with 10 mL mineral medium in sterile 30 mL glass serum bottles which were sealed under an atmosphere of 70% N$_2$, 28% CO$_2$, and 2% H$_2$ in a glove box. Soils were brought to different redox-status to investigate the anaerobic biodegradability under denitrifying, iron(III)-reducing, sulfate-reducing, and methanogenic conditions for the sand soil; loamy sand was tested under denitrifying conditions, only. Soils were spiked with about 3 ppm radiolabeled 4,4′-MDA. At the end of the tests, aqueous phase radioactivity (aqueous phase filtered through a 0.7-μm membrane), total extractable radioactivity, and particle bound radioactivity were measured. For extractable radioactivity, 1 mL 1 N NaOH solution was added to the test mixtures, followed by 1 mL of acetonitrile so it made up 7% of the mixture, and after mixing for 1 h, radioactivity was measured in the supernatant after centrifugation and filtration through a 0.7-μm filter. For particle bound radioactivity, combustion analysis of the solid pellet gained after centrifugation was performed. The centrifuged and filtered extraction samples were also subjected to HPLC-radiodetection analysis. The extraction procedure was found to be most efficient with an MDA recovery of 68%.

In loamy sand under denitrifying conditions, particle bound radioactivity was $44 \pm 7\%$, total extractable radioactivity was 7%, and dissolved organic matter (DOM) bonded radioactivity was 18% after 374 days; in killed controls, 57% of radioactivity was particle bound, 4% was DOM bound. In sand soil under denitrifying conditions, $54 \pm 6\%$ radioactivity was particle bound, extractable radioactivity was $27 \pm 5\%$, the latter bound to DOM; in killed controls, $28 \pm 4\%$ radioactivity was particle bound and $27 \pm 2\%$ radioactivity was extractable. When the denitrifying soils were reduced by addition of volatile fatty acids (VFA), 77% radioactivity was recovered from the aqueous phase after 1 year, being composed of MDA (12%) and a metabolite (65%) which was most likely 4,4′-diaminobenzophenone; 24% MDA was reversibly adsorbed to the soil. Obviously, no MDA was bound to DOM (West et al. 2002).

In sand soil under iron(III)-reducing conditions, the aqueous phase contained 23% radioactivity, and the total extractable radioactivity was more than 90%, made up mostly by MDA and about 10% of unidentified degradation products; MDA seemed not to be bound to DOM. Killed controls delivered the same results as the killed denitrifying aquifer sand batches (West et al. 2002).

Aquifer sand soil under sulfate-reducing and methanogenic conditions, total extractable radioactivity after 374 days was more than 90%, almost all identifiable as MDA (West et al. 2002). In biologically inhibited batches, MDA made up only 4% of the extractable radioactivity.

4.2 Studies on Biodegradation

Screening tests on ready and inherent biodegradability are summarized in Sects. 3.1 and 3.2, respectively.

In degradation tests with radiolabeled MDA in sandy loam and silt loam under aerobic conditions, CO_2 was formed continuously over the observation period of 365 days (Cowen et al. 1996, 1998). The first-order rate constants for the mineralization can be calculated by:

$$\frac{1}{t} \times \ln \left(\frac{C_0}{C} \right) = k;$$

In this formula, t is the respective time interval Δt between data points; the concentration C is calculated as $(100 - \% \, CO_2)/100$ for the respective data point. The immediately starting biodegradation of the MDA slowed down, and irreversible adsorption to soil can explain this finding. Data for CO_2 yields and first-order rate constants for MDA mineralization are given in Table 5. Initially, the formation of CO_2 is comparatively rapid; however, the radiochemical purity of the MDA was 98%, so impurities (probably aniline) are expected to have contributed to the initial CO_2 yield. Over time, the mineralization rate constant varies around 3.6×10^{-4} days^{-1} to 1.38×10^{-3} days^{-1}. This observation is remarkable in so far as for another primary

Table 5 Yields of CO_2 and first-order mineralization rate constant for MDA in soil under aerobic conditions (Cowen et al. 1996)

Day	Total CO_2 (%)	ΔCO_2 (%)	Δt (days)	$k \times 10^3$ (1/day)
0	0	0	0	–
1	2	2	1	20.20
3	3	1	2	5.03
7	9	6	4	15.47
14	10	1	7	1.44
28	11	1	14	0.72
56	12	1	28	0.36
210	31	19	154	1.37
365	40	9	155	0.61

aromatic amine, 3,4-dichloroaniline, Saxena and Bartha (1983) observed a similar behavior. They showed that the mineralization rate of 3,4-dichloroaniline bound by humic acids aligns with the mineralization rate of natural organic matter.

Cowen et al. (1996) also checked the anaerobic biodegradation of MDA. Under the methanogenic anaerobic conditions chosen, neither radiolabeled CO_2 nor CH_4 was detectable after 365 days.

Kim et al. (2002) investigated the biodegradation of MDA in the modified Sturm Test (OECD 301B). As inoculum, they took activated sludge soil from soil around an aeration tank of a wastewater treatment plant. At a loading of 30 mg/L MDA, they observed a yield of CO_2 of 74% after 35 days and about 60% after 28 days; as the 10-day window was not passed, MDA cannot be described as readily biodegradable. Higher concentrations of MDA resulted in poorer biodegradation, indicating some toxicity of the MDA to the inoculum. The latter was taken from soil around STPs where digester sludge had been disposed of. They also isolated specific MDA degrading organisms, *Ochrobactrum anthropi* and *Aspergillus* sp., typical soil species, which showed comparable rates of MDA mineralization and also similar inhibition by increased MDA concentrations.

West et al. (2002) performed an extended anaerobic biodegradation study with radiolabeled MDA in either aquifer sand or loamy sand. 25 g soil (dry weight) were mixed with 10 mL mineral medium in sterile 30 mL glass serum bottles which were sealed under an atmosphere of 70% N_2, 28% CO_2, and 2% H_2 in a glove box. Soils were brought to different redox-status to investigate the anaerobic biodegradability under denitrifying, iron(III)-reducing, sulfate-reducing, and methanogenic conditions. Soils were spiked with about 3 ppm radiolabeled 4,4'-MDA, and the yield of labeled CH_4, CO_2 as well as extractable radioactivity was measured over time.

Under denitrifying condition, the half-life time for extractable MDA was biphasic with <0.1 days and 94 days in loamy sand and was 34 days in aquifer sand. After 1 year, 0.2 and 1.7% radiolabeled CO_2 were generated in loamy sand and aquifer sand, respectively. In biologically inhibited denitrifying soils (autoclaving and addition of $HgCl_2$), the half-life of MDA in sand soil dropped to 12 days; the yield of CO_2 was 0.2% at the end of the test period for both soils under denitrifying conditions. When p-aminobenzoate was added to the aquifer sand batch, the yield of CO_2 increased from 1.7 ± 0.7 to $2.7 \pm 0.5\%$, whereas the half-life time for MDA absorption (decay) increased from 34 to 50 days; therefore, p-aminobenzoate facilitates MDA biodegradation either as co-substrate, or it keeps MDA longer available for the bacteria by blocking binding sites in the soil. In loamy sand under denitrifying conditions, CO_2 yield was 0.2% in viable microcosms as well as in sterilized controls. Nitrate dropped down to 40% of the initial value, and traces of nitrogen oxides were detectable in the headspace. In killed controls, no nitrate conversion was observable. When the batches under denitrifying conditions were poised with VFA, the half-life of MDA extended 136 days (West et al. 2002).

Yields of labeled CO_2 under Fe(III)-reducing conditions, tested in aquifer sand only, were $0.7 \pm 0.7\%$; in the killed controls, the yield of labeled CO_2 was 0.6% (West et al. 2002).

For sulfate-reducing conditions (aquifer sand only), the CO_2 yield was $0.7 \pm 0.4\%$, and 0.1% in killed controls (West et al. 2002).

Anaerobic degradation under methanogenic conditions for aquifer sand delivered $0.5 \pm 0.3\%$ CO_2, and 0.1% in killed controls (West et al. 2002).

To minimize the irreversible binding of MDA to soil humic matter in anaerobic biodegradation tests, West performed tests in anaerobic enrichment cultures (2007). Enrichment cultures were generated starting from anaerobic aquifer sand and loamy sand as used by Cowen et al. (1996). The inocula were diluted in anaerobic mineral medium, brought to denitrifying, iron(III)-reducing, sulfate-reducing, or methanogenic conditions and fed with VFA in two portions over 37 days; addition of VFA minimizes irreversible absorption of MDA to soil (West et al. 2002). Then, 5 mg/L radiolabeled MDA was added and its transformation was followed for up to 865 days. The cultures contained finally about 200 mg soil per liter. Labeled acetate and aniline served as controls, and aniline was used for the toxicity control. Under denitrifying conditions, about 3% 4,4'-diaminobenzophenone was formed in the culture derived from the aquifer sand. The formation of labeled CO_2 and CH_4 was below 1%, respectively; under sulfate-reducing conditions, in two experiments no labeled CO_2 was generated, whereas in the third batch yield of CO_2 was 34%. For Fe(III)-reducing conditions, the sum of labeled CO_2 and CH_4 was below 1% after 865 days. Under methanogenic conditions, about $2.1 \pm 2.3\%$ CO_2 and CH_4 were formed. Analysis revealed that over the testing period nitrate and sulfate were consumed under denitrifying and sulfate-reducing conditions, respectively. Iron (III) was reduced to iron(II), and nitrate and sulfate were consumed in the relevant viable cultures, but not in killed controls. The presence of MDA retarded the degradation of aniline under sulfate-reducing conditions. It is not clear whether this retardation was attributable to toxicity of MDA to the bacteria or whether it was simply due to competition between MDA and aniline, as the highest mineralization rate for MDA was observed under sulfate-reducing conditions. The decay of MDA in the batches followed first-order kinetics with a rate constant between

Table 6 Results of degradation tests with MDA in soil environments and rate constants for mineralization (k_{min}) and primary degradation (k_{prim})

Test	Result	Remarks	Reference
Soil, aerobic[a]	40%, CO_2	CO_2 yield after 365 days	Cowen et al. (1996, 1998)
Soil, anaerobic methanogenic[a]	0% CO_2, 0% CH_4	72 days test duration	Cowen et al. (1996, 1998)
Soil, anaerobic denitrifying[a]	1.7% CO_2	374 days test duration; biphasic decay of MDA	West et al. (2002)
Soil, anaerobic methanogenic[a]	0.5% CO_2	374 days test duration	West et al. (2002)
Enrichment culture, anaerobic, methanogenic[a]	2.1 ± 2.3% CO_2 and CH_4	865 days test duration	West (2007)
OECD 301B	60% CO_2	Soil inoculum	Kim et al. (2002)

[a]Redox-status of the soil or culture, resp.

0.001 and 0.036 days^{-1} and was rather uniform in biotic and abiotic controls. This study showed that MDA can be mineralized under anaerobic conditions, but abiotic decay processes are more important for the disappearance of MDA. Either the limited amount of soil still represented a sufficient number of reactive sites or other, soil-independent reaction pathways play an important role. Data concerning MDA decay in soil are summarized in Table 6.

4.3 Combined View: Behavior of 4,4'-Methylenedianiline in Soil

As indicated in Sect. 4.1, MDA is obviously capable to form non-extractable residues with soil organic matter under aerobic conditions and under anaerobic, denitrifying conditions (Cowen et al. 1996; West et al. 2002). Due to this fact, investigations concerning distribution in soil do not actually deliver equilibrium constants, unless tests are performed in anoxic, highly reduced soil. Therefore, use of K_{oc} or K_d for the interpretation of the environmental behavior of MDA has limited meaning. MDA can be mineralized in aerobic soil, and the mineralization rate approaches levels typical for soil organic matter (Saxena and Bartha 1983). Under anaerobic conditions, mineralization becomes very poor, and yields of CO_2 and CH_4 have to be interpreted with caution as the radiochemical purity of the MDA was 98%. The primary decay of MDA is rapid under aerobic and anaerobic, denitrifying conditions but becomes poorer in soils with lower redox potentials. West et al. (2002) argue that under viable and active anaerobic conditions the redox potential of the soil would be below that of quinone/hydroquinone, shifting the equilibrium towards hydroquinone. Quinoic structures are a known part of humic matter and dissolved organic carbon (West et al. 2002). Quinone, but not hydroquinone, can

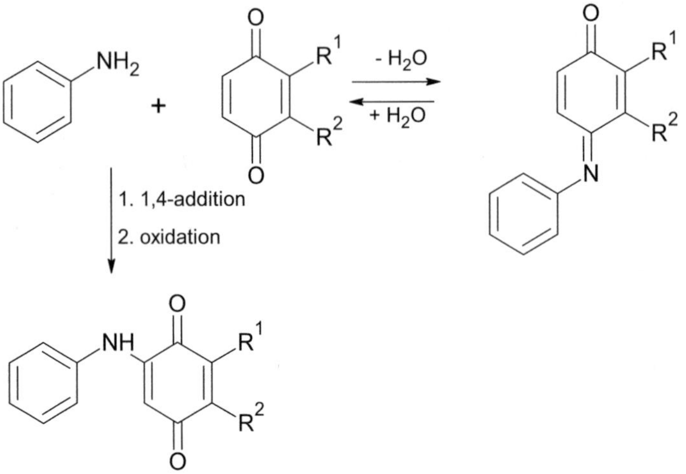

Fig. 4 Reaction of aniline with quinones in humic matter

react with primary aromatic amines under formation of covalent N–C bonds, which explains the irreversibility of the absorption (Parris 1980; Weber et al. 1996, 2002). The reaction is illustrated in Fig. 4. West et al. (2002) also described the formation of MDA metabolites under anaerobic regimes in soil. These were tentatively identified by HPLC-MS. 4,4′-Diaminobenzophenone was always formed, but predominantly under denitrifying conditions, and a maximum yield of 65% could be achieved in VFA amended, oxygen poised batches under denitrifying conditions. Another metabolite was p-aminobenzaldehyde, which achieved a yield of 10–13% and was observable in killed controls, only. As the killed controls were generated by addition of mercuric chloride, it is not clear whether this metabolite was detectable due to blocked microbial degradation, or whether it was a result of direct reaction between MDA and mercury. West et al. (2002) concluded that in principle MDA can be degraded and mineralized under denitrifying anaerobic conditions, but that irreversible absorption to soil humic matter traps the MDA so it is no longer available for the bacteria. The lower the redox-status of the soil, the less likely the irreversible adsorption of MDA will occur.

Taken together, for the behavior of MDA in soil kinetic approaches might be favorable equilibria relations. The possible interaction of adsorption, absorption, metabolism, and mineralization can be described as shown in Fig. 5. Once the kinetic rate constants are available, a semiquantitative evaluation and derivation of steady-state equilibria might be possible.

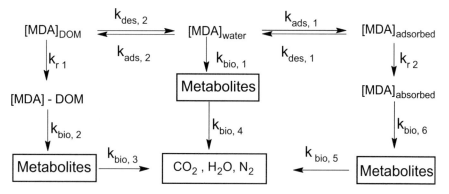

Fig. 5 Simplified model of MDA pathways in soil; k = first-order rate constants for adsorption, desorption, reaction, and biotransformation

5 Behavior in the Atmosphere

First calculations based solely on physical properties predicted that air plays a minor role for the distribution of MDA. Nevertheless, data for the primary photodegradation of MDA in air are available (Becker et al. 1988). Reactions were performed in a Duran glass chamber at 10^5 Pa and 298 K, and concentrations were followed by FT-IR spectroscopy. OH radicals were generated by photolysis of methyl nitrite, and added NO trapped any O_3 and NO_3*, so decay of MDA should be attributable to reactions with OH* only. Experiments in the dark or in the absence of OH* delivered data for the non-photolytic disappearance of MDA, which was described as reaction with the chamber wall. In the presence of OH*, the decay was always discernible faster than without OH*. The rate constant was $3 \pm 1 \times 10^{-11}$ cm^3/s. Therefore, the half-life time is 6.4 h when the concentration of OH* is 10^6 molecules/cm^3.

Pemberton and Tury (2008) calculated a rate constant for the photodegradation of MDA of 2×10^{-10} cm^3/s by use of the AOPWIN™ program. The corresponding half-life time in air is 1.2 h when the daily average OH* concentration is 1×10^6 molecules/cm^3. This value is somewhat smaller than the measured decay rate. In both cases, MDA suffers a comparatively fast decay in air. MDA is a substrate for oxidants, as was revealed in the tests for degradation in soil, and, therefore, should also be a substrate for reactions with ozone, singlet oxygen, and other reactive oxygen species. An atmospheric decay rate constant of 3×10^{-11} cm^3/s is most likely the lower bound value for atmospheric reactions.

6 Environmental Degradation Products of 4,4′-Methylenedianiline

Degradation of the MDA substances by the various processes described above may result in formation of degradation products which should themselves be assessed for environmental persistence, bioaccumulation potential, and inherent toxicity. The OECD screening tests of ready and inherent biodegradability do not require identification of intermediate degradation products; however, some evidence such as the observation of yellow color formation in the tests of Mei et al. (2015) indicate that products intermediate of the parent MDA and carbon dioxide end product are formed which have some stability in the test medium. The studies by Schaefer and Carpenter (2013) and Schaefer and Ponizovsky (2013) showed that several radiolabeled degradation products were detected in both aerobic and anaerobic surface waters and associated sediments. Biologically active systems produced five extractable metabolites being probably more polar than MDA (shorter retention times in reversed-phase HPLC runs) and three metabolites less polar than MDA (higher retention times); in killed controls, only two metabolites were detected (Schaefer and Carpenter 2013). However, the very low initial concentrations of [^{14}C]-MDA used in these studies did not allow for isolation and identification of these degradation products. More recently, a study was conducted where these test conditions were replicated (from the same river water/sediment collection site) in an attempt to form and concentrate these same degradation products and enable tentative identifications by HPLC-mass spectrometry (Campbell 2017). Three such transformation products (and associated isomers) which were tentatively identified indicated that a number of oxidative and reductive transformations of MDA had occurred in the aerobic water/sediment mixtures. These included sequential oxidations would result in products having hydroxyl and ketone functional groups at the methylene carbon ($-CH_2-$) which bridges the aromatic rings of MDA. Other suspected transformations involved multiple N-acetylation, N-sulfation, N-sulfonation, and N-phosphorylation transformations at both primary amino groups. Some structures of the proposed transformation products showed a combination of these transformations with reduction of the aromatic ring to a cyclohexyl moiety, while others showed hydroxylation of the aromatic rings at unspecified positions. Because the total mass of transformation products isolated was low, MS–MS fragmentation experiments which could allow elucidation of structure sub-fragments and positions of ring substituents did not produce useful information. Therefore, the assignments of transformation product identities made in this study for MDA are tentative. However, hydroxylation of the methylene bridge and of the aromatic ring would result in metabolites more polar than MDA, whereas N-acetylation would generate metabolites less polar than MDA.

A look on metabolism studies with mammals may help to understand what kind of metabolites of MDA can be found. Metabolism of MDA was studied in rats (Morgott 1984). After i.p. administration, about 17 metabolites were found in the urine. The main metabolites were N-acetyl-MDA; N,N′-diacetyl-MDA; N,N′-diacetyl-3-hydroxy-MDA;

N,N'-diacetyl-diaminobenzophenone; and N,N'-diacetyl-diaminobenzhydrol. With rabbit liver microsomes, metabolites found in vitro were azo-MDA, azoxy-MDA, and 4-nitroso-4′-aminodiphenylmethane (Kajbaf et al. 1992).

Considering the paucity of information coming from direct detection and structural elucidation of MDA degradation products from environment simulation tests, perhaps the best insights into their identity comes from libraries of metabolic pathways for structurally related substances, and from the metabolite prediction algorithms which are built from them. The Biodegradation and Catalysis Database [originally compiled by the University of Minnesota and now curated and expanded by the Swiss Federal Institute of Environmental Science and Technology (EAWAG)] shows numerous aerobic microbial catabolic pathways for primary aromatic amine substances (http://eawag-bbd.ethz.ch/index.html). Their web site provides a microbial metabolism pathway prediction system, which gives relative probabilities for various pathways and associated metabolites for an input organic chemical, based on the metabolic pathway library. For MDA, the most probably predicted biodegradation pathways involve de-amination of one ring with concomitant hydroxylation to form a (4-aminophenyl)-catechol intermediate. The catechol ring is cleaved to give various isomers of 4-aminophenyl-muconic acid derivatives, which are further mineralized through a common 4-aminobenzoic acid intermediate. The database includes known pathways for 4-aminobenzoate, and numerous other primary aminobenzene analogs, which are degraded by the same pathway of de-amination and subsequent catechol formation. Concerning experiments with MDA, hydroxylation of the aromatic ring was indicated (Campbell 2017), but deamination was not observed, so far.

Another pathway prediction capability is provided by the Laboratory of Mathematical Chemistry at the University Prof. Dr. Asen Zlatarov, Bourgas, Bulgaria (http://oasis-lmc.org/). Their OASIS Catalogic software package also provides relative probabilities of various predicted pathways and metabolites, which are based on algorithms different from, but including pathways of the EAWAG-BBD and other pathway libraries (Karabunarliev et al. 2012). This software predicts MDA degradation pathways which involve de-amination and dihydroxylation of the aromatic rings, as well as oxidation of the bridging methylene carbon to give amino benzophenone derivatives (see Annex 1). The latter have been observed in the studies of West et al. (2002) and West (2007) and in other studies of environmental degradation with the analog methylenebis-(2-chloroaniline) (Voorman and Penner 1986a). The abiotic oxidative coupling of primary aromatic amines can occur in the presence of molecular oxygen and various metal catalysts to form aromatic azo-linked dimer substances (Burge 1972; Briggs and Olgivie 1971; Kaufman et al. 1973; Pillai et al. 1982). While some studies have implicated biological activity and associated peroxidase enzymes in this reaction, the reactions have been demonstrated in soils which were sterilized by gamma irradiation and were presumed to be coupled with reduction of manganese and/or iron species in the soil (Li et al. 2003).

In summary, the observed and expected pathways for microbial metabolism of MDA indicate oxidation of the methylene bridge, and probably also hydroxylation of the aromatic ring. These transformations would produce more polar compounds than MDA; N-hydroxylation would result in less polar metabolites; more polar and less polar metabolites have been observed by Schaefer and Carpenter (2013), and only two of five more polar metabolites may be attributable to abiotic transformations as these were observed in killed controls as well. The OASIS program predicts transformations which serve to remove (i.e., oxidative de-amination) or reactively substitute (e.g., azo coupling, N-acetylation, etc.) the primary aromatic amine functionality to which toxicity of the substance is attributed. However, the experimental proof for MDA deamination is lacking.

7 Accumulation in Biota and Metabolism

In accumulation studies in carp (*Cyprinus carpio*) according to OECD Guideline no. 305, the bioconcentration factor was 3–14 at 200 μg/L MDA in water, and <3–15 at 20 μg/L (CITI 1992).

Metabolism of MDA was studied in rats (Morgott 1984). After i.p. administration, about 17 metabolites were found in the urine. The main metabolites were N-acetyl-MDA; N,N'-diacetyl-MDA; *N,N'*-diacetyl-3-hydroxy-MDA; *N,N'*-diacetyl-diaminobenzophenone; and *N,N'*-diacetyl-diaminobenzhydrol. After 3 days, more than 90% of the applied MDA was excreted with urine and feces. With rabbit liver microsomes, metabolites found in vitro were azo-MDA, azoxy-MDA, and 4-nitroso-4'-aminodiphenylmethane (Kajbaf et al. 1992).

Monkeys dosed intravenously with MDA excreted more than 90% of the material within 48 h (ECHA 2014a).

Data for MDA metabolism in fish tissue are not available. However, qualitatively trout is comparable to rat concerning metabolic competence, though the activities of the enzymes are at a lower level (Nabb et al. 2006). For example, CYP450 monooxygenases in trout liver cells achieve 5–30% of the activity of rat hepatocytes, for chloro-dinitrobenzene conjugation to glutathione the activity is 7% compared to rat. For the clearance of N,N,N',N'-tetramethyl-4,4'-diaminobenzophenone, Han et al. (2007) report a hepatic clearance of 156 mL/h/kg and a first-order rate constant of 0.671 day^{-1}.

Tomato plants were exposed to MDA-spiked water in hydroponic cultures for 5 days at pH = 5.5 and 6.5 (Bongartz 2012). Given the pKa values in Sect. 2 (Physical–Chemical Properties), the ratio of MDA to $MDAH_2^{2+}$ is 50 at pH = 5.5 and 500 at pH = 6.5; the ratio of MDA to MDAH$^+$ is 2 at pH = 5.5, and 20 at pH = 6.5. The radiochemical purity of the MDA was 95.84%, and there were two unknown impurities, one with lower retention time than MDA in reversed-phase HPLC (1.56%) and one with higher retention time than MDA (2.59%). Hydroponic cultures were chosen to ensure that MDA would not be trapped by soil components. Within the first 4 h of the test, there was a strong decay of MDA in solution (40–50%) which was interpreted as adsorption onto the roots. After 5 days, the radioactivity in solution was about 10% of the initial activity. Without plants, there was only a very slow decay of radioactivity in the solution, and after 5 days it was more than 90% of the initial value (blank experiment). The plant uptake factor was about 7 at pH = 5.5 and about 8 at pH = 6.5. Radiography showed that the activity in the plants was mainly at the outer layers of the roots, virtually none in the stem and very weak activity in the leaves. Combustion radiography resulted in 5.91 mg/kg MDA in roots and 0.056 mg/kg MDA in leaves, if the radioactivity was attributable to MDA. The samples were too small and the dilution too high for specific analysis.

8 Environmental Distribution and Long-Range Transport

The distribution and fate of MDA was modeled with the EUSES program v.2.1 (European Union 2016). Decay of MDA can be modeled as primary degradation and as mineralization. With regard to the data generated, conditions may be favorable (high mineralization rates, primary degradation of MDA, and low accumulation constants) and unfavorable (low mineralization rates, negligence of primary degradation, and high accumulation constants). The negligence of primary degradation under the "unfavorable" setting is owed to the fact that disappearance of MDA is not necessarily a detoxification as long as the metabolites are not identified and proven to be harmless.

For the production volume, the EU risk assessment (European Union 2001) mentions a production volume of 432,000 t/a; for convenience, we assume a current production volume of 500,000 t/a, and a default release of 0.3% into waste water. This assumption overestimates the release (European Union 2001), but the purpose of this modeling now is not a site- and region-specific risk assessment but an overview of the influence of degradation pathways on the distribution of MDA. Results are given in Table 7.

With an atmospheric half-life time of less than 2 days, MDA is not suspected to undergo long-range transport.

The influence of the degradation constants on the predicted environmental concentrations (PEC) is most obvious for the soil and sediment compartments. The very high degradation rate constants for surface water and sediment are attributable to primary decay. For soil, only rate constants derived from CO_2 formation, that means mineralization, are used. Although data on primary and inherent biodegradation of MDA delivered values of "not inherently biodegradable" to "inherently biodegradable" to "readily biodegradable," depending on the study (see Sect. 3), the use of 0.1 and 0.3 h^{-1} for first-order rate constants in the EUSES program seems justified for the following reason: as outlined in the European Risk Assessment report (European Union 2001), release into the environment is expected on industrial sites only; the receiving compartment is the surface water, after the substance has passed the industrial wastewater treatment plant; experimental data show that it is very likely that microorganisms can be adapted to the digestion of MDA. Regional sediment, followed by agricultural soil and finally industrial soil are the compartments with highest values for the predicted environmental concentration of 4 to 8 ppm. For the full EUSES output, see Annex 2.

Table 7 EUSES 2.1 calculation results

Section/parameter	Favorable	Unfavorable
Rate constant for biodegradation in STP (h^{-1})	0.3^a	0.1^a
Rate constant for photolysis in surface water (day^{-1})	$3.6E-04^b$	$3.6E-04^b$
Rate constant for biodegradation in surface water $(day^{-1}, 12°C)$	0.0924^b	$2.16E-03^a$
Rate constant for biodegradation in aerated sediment $(day^{-1}, 12°C)$	0.224^b	$2.84E-04^a$
Rate constant for degradation in air (day^{-1})	0.7^b	0.7^b
Rate constant for biodegradation in bulk soil $(day^{-1}, 20°C)$	$5.11E-03^a$	0^a
Regional PEC in surface water (total) (mg/L)	3.42E-03	0.0289
Regional PEC in seawater (total) (mg/L)	3.3E-04	2.79E-03
Regional PEC in surface water (dissolved) (mg/L)	3.38E-03	0.0286
Regional PEC in seawater (dissolved) (mg/L)	3.29E-04	2.79E-03
Regional PEC in air (total) (mg/m^3)	2.02E-09	2.04E-09
Regional PEC in agricultural soil (total) (mg/kg wwt)	0.0557	7.1
Regional PEC in pore water of agricultural soils (mg/L)	4.51E-04	0.0574
Regional PEC in natural soil (total) (mg/kg wwt)	1.42E-03	0.0428
Regional PEC in industrial soil (total) (mg/kg wwt)	0.146	4.38
Regional PEC in sediment (total) (mg/kg wwt)	0.111	7.98
Regional PEC in seawater sediment (total) (mg/kg wwt)	4.48E-03	0.684

[a]Based on mineralization
[b]Based on primary decay

9   PBT Assessment

West and Tury (2006) analyzed available data for MDA against the criteria for classification of substances as persistent, bioaccumulative, and toxic (PBT). Newly generated data after 2006 are collected and analyzed for PBT assessment. Criteria for the classification of chemicals as persistent (P), bioaccumulative (B), and toxic (T) are defined in different regions. PBT classification is judged on a yes/no decision by legislatively set cutoff values. As an example, the criteria for the European Union (EU) and the United States Environmental Protection Agency (US EPA) are summarized in Table 8.

Both of ready biodegradation tests OECD 301A and OECD 301F (Mei et al. 2015) indicate that the substance is readily biodegradable. Simulation test OECD 308 indicates DT50 of 5.8 days for estuarine water sediment and 3.1 days for freshwater sediment under aerobic conditions (Schaefer and Ponizovsky 2013), and 21 days for estuarine water sediment, and 10 days for freshwater sediment under anaerobic conditions. Simulation test OECD 309 indicates DT50 of 11 days for estuarine water and 7.5 days for freshwater (Schaefer and Ponizovsky 2013).

Based on the DT_{50} values and half-lives, MDA should not be regarded as persistent against current PBT criteria of the EU, the USA, and Canada; MDA should also not be regarded as persistent under the criteria of the UNEP Stockholm POPs and OSPAR conventions. However, the ECHA stresses that metabolites and bound residues need to be considered if the half-lives are not based on mineralization. Taken all degradation

Table 8 PBT criteria

	Persistence (half-life)	Bioaccumulation potential	Toxicity
European Chemicals Agency	**P**: $T_{1/2} > 60$ days in marine water, or >40 days in fresh or estuarine water, or >180 days in marine sediment or >120 days in soil or fresh or estuarine water sediment; **vP**: $T_{1/2} > 60$ days in water or >180 days in sediment or soil	BCF > 2000: **B** BCF > 5,000: **vB**	NOEC or EC_{10} (long-term) <0.01 mg/L; or carcinogen or mutagen category 1 or toxic to repro-duction; or other evidence
US EPA PBT control action plan	Air >2 days; aquatic environment >60 days	BCF > 1,000	Develop data where necessary, based on concerns for P, B, or other factors
US EPA PBT ban action plan	Air >2 days; aquatic environment >180 days	BCF > 5,000	Develop data where necessary, based on concerns for P, B, or other factors
UNEP Stockholm POPs convention	Air >2 days; water >60 days; sediment, soil >160 days or "sufficient concerns"	BCF > 5,000 or BAF > 5,000 or Log Kow > 5.0	Expert judgment
Oslo-Paris convention for the protection of marine environment of the north-east Atlantic (OSPAR)	Not readily biodegrad-able or $T_{1/2} > 50$ days in simulation tests	BCF > 500 or Log Kow > 4	Acute aquatic L(E) $C_{50} < 1$ mg/L or chronic aquatic NOEC <0.1 mg/L or human health carcinogen, muta-gen or toxic to reproduction
Environment Canada	Air ≥2 days Water ≥182 days Sediment ≥365 days Soil ≥182 days	BAF ≥ 5,000 or BCF ≥ 5,000 or Log Kow ≥ 5.0	Acute aquatic L(E) $C_{50} < 1$ mg/L or chronic aquatic NOEC <0.1 mg/L or human health carcinogen, muta-gen or toxic to reproduction
Japan Chemical Substance Control Law	Fail to pass OECD 301C, 302C (MITI biodegradation tests) And/or environmental monitoring (for existing chemical)	BCF > 5,000 is "B" BCF < 1,000 or log Pow < 3.5 is not "B"; 1,000 < BCF < 5,000, expert judgment	Evidence of long-term effects to humans or environment

information together, MDA can be mineralized by microorganisms, but irreversible binding to organic matter competes with the biodegradation. MDA is likely to be incorporated into natural organic matter. Like for dichloroaniline, based on biodegradation tests in aerobic soil (Cowen et al. 1996) it can be assumed that after binding to organic matter, the rate of degradation of MDA moieties will then be determined by the rate of natural organic matter decay (Saxena and Bartha 1983). For Japan, MDA fails to pass the OECD 301C and 302C tests. However, in Japan National Institute of Technology and Evaluation's MDA risk assessment report (NITE 2007), it is mentioned that MDA is not "persistent" referring to the report of Howard et al. (1991), which reported that MDA can be degraded by activated sludge. In anaerobic environments, however, mineralization of MDA becomes negligible. Metabolites are formed, but MDA is more and more left unchanged with lower redox potential of the soil matrix; irreversible absorption takes place under anaerobic conditions in soil, but in highly reduced soil, irreversible absorption and formation of non-extractable residues becomes negligible (West et al. 2002). Therefore, MDA and/or its metabolites may be persistent in anaerobic, highly reduced environments.

MDA is not bioaccumulative in any of the evaluation schemes listed since its log Pow is 1.55 and the BCF is ≤ 15 (CITI 1992). In mammals, MDA undergoes phase-I and phase-II metabolism finally resulting in partly more polar compounds than MDA, and partly less polar compounds (Morgott 1984; Kajbaf et al. 1992). The less polar compounds are N-acetylated MDA and azo-bis(4-(4-aminobenzyl)benzene. Their calculated log K_{ow} values are 2.33 and 7.04 at pH $= 7.5$, respectively (SPARC calculator, ARChem LLC, 2017; http://archemcalc.com/sparc-web/calc.). The azo-compound should be regarded as bioaccumulative due to this calculated log K_{ow}. However, in monkeys dosed intravenously with MDA, more than 90% of the material was excreted within 48 h (ECHA 2014a), and rats excreted more than 90% of the dosed MDA within 3 days (Morgott 1984). These experimental findings do not indicate that MDA is bioaccumulative in mammals. Data concerning behavior in invertebrates are not available. As MDA is rapidly adsorbed to plant roots (Bongartz 2012), in the absence of the identification of the adsorbed material the question of secondary poisoning cannot be answered, yet.

MDA is toxic, as it is a carcinogen category 1b and a carcinogen in chronic tests with mice and rats (NTP 2018). Ecotoxicity data are summarized in Schupp et al. (2016). Briefly, the most sensitive aquatic species is *Daphnia magna* with an EC_{50} (48 h) of 0.35 mg/L and an 21 d NOEC of 0.005 mg/L; the most sensitive sediment species is *Lumbriculus variegatus* with a 28-d NOAEC of 3–30 mg/L; lowest effect levels for soil organisms were the 56-d EC_{10} of 11 mg/kg for *Eisenia fetida*.

As a conclusion, MDA is not bioaccumulative; it is not persistent in terms of primary decay; under certain unfavorable circumstances, MDA itself or its metabolites may be persistent, and under favorable conditions it may show a considerable rate of mineralization; MDA is very toxic to aquatic life; taken together, MDA should not be classified as PBT or vPvB substance.

10   Conclusion

According to most of the OECD Guideline 301-series reported, MDA is not readily biodegradable. Results in these tests series are variable. While degradation was virtually 0% in an OECD 301C test (CITI 1992), in an OECD 301B test, CO_2 achieved 46% after 28 days, and 53% after 63 days (Schwarz 2009). Mei et al. (2015) reported ready biodegradability in OECD 301A and 301F, but not in 301B and 301D tests. In their argumentation, they hypothesize that incorporation of MDA in biomolecules could explain the high DOC removal as well as the high oxygen consumption, while the formation of carbon dioxide is limited. MDA may give reason to abiotic oxygen consumption, with 4,4'-diaminobenzophenone and/or azo-compounds as potential products; however, these oxidations would consume only 9.1% of the oxygen required for full mineralization. The different results in the tests for ready or inherent biodegradability may be attributable to the statistical presence/absence of specific bacteria, in addition to the general variability of these tests, and the fact that at elevated concentrations, MDA inhibits microorganisms (Kim et al. 2002). In an OECD 302C tests, mineralization of MDA was 43% (Caspers et al. 1986), whereas in an OECD 302B test MDA was nearly completely removed (BASF 1981). In the latter test, inoculum from an industrial STP was used, and it is likely that microorganisms were adapted to MDA. Under anaerobic conditions in the OECD 308 test, in killed controls DT_{50} values were lower than in viable samples. These results indicate that anaerobic microbial activity results in reduced redox potentials which either remove MDA reactive quinoic structures, and/or block the oxidation of MDA, as discussed below.

In an OECD 301B test with soil inoculum, MDA degradation achieved 60% CO_2 evolution, but missed the 10-d window (Kim et al. 2002); by use of enrichment cultures and selective media, the authors were able to isolate MDA degrading microorganisms. In soil, MDA mineralization starts immediately but slows down rapidly (Cowen et al. 1996). First-order mineralization rate constants of initially 2.02×10^{-2} days^{-1} reduce to about 1.0×10^{-3} days^{-1}, which is about the mineralization rate of soil organic matter (Saxena and Bartha 1983). Irreversible absorption seems to compete with biodegradation (see below).

In anaerobic soils, mineralization becomes negligible (Cowen et al. 1996; West et al. 2002). Nevertheless, a primary decay of MDA is observable unless in anaerobic, highly reduced soils. This indicates that reaction with quinoic structures in soil organic matter plays an important role. MDA also forms some extractable metabolites under these conditions, one of them most likely being 4,4'-diaminobenzophenone; this is one main metabolite formed under denitrifying conditions.

Concerning degradation in soil, irreversible absorption, probably by covalent binding of MDA to soil organic matter seems to compete with its mineralization. One important side reaction is the reaction with humic substances under formation of covalent bonds. Similar behavior was observed with the structural analogue of MDA, methylene-3,3'-dichloro-4,4'-dianiline (methylene-4,4'-bis-(ortho-chloroaniline), MBOCA), as reported by Voorman and Penner (1986a). 4-chloroaniline,

3,4-dichloroaniline and 2,6-diethylaniline react with the lignin and humic matter model compounds vanillic acid, syringic acid, protocatechuic acid, and ferulic acid under oligomerization (Bollag et al. 1983, 1992). These reactions were catalyzed by laccase, an enzyme present and excreted by diverse rot fungi. Covalent bonds were formed between carbon and nitrogen. As MDA is a bifunctional aromatic amine, its reaction with humic matter is expected to result in oligomers or even polymers. Binding to humic matter does not necessarily prevent further biodegradation of the aromatic amine, as demonstrated by Hsu and Bartha (1974) for 3,4-dichloroaniline. For 3,4-dichloroaniline, binding to soil humic matter reduces the mineralization rate of that amine to a level equivalent to humic matter (Saxena and Bartha 1983). A comparable behavior is expected for MDA.

In the presence of soil and sediment microbial activity, MDA forms metabolites; some are more polar than MDA, some are less polar according to retention times in reversed-phase HPLC runs; in killed controls, much fewer metabolites are formed (Schaefer and Carpenter 2013). Oxidation of the methylene bridge is a likely metabolism pathway. HPLC-analysis identified fragments indicating this methylene bridge oxidation, but also hydroxylation of the aromatic ring, N-acetylation, N-sulfation, and N-phosphorylation. Such fragments are predicted by the OASIS program; deamination as predicted by the OASIS program was not yet experimentally proved to happen with MDA.

Measured 8-h log K_{oc} values of about 3.6–3.9 from experiments with uncertain pH are in agreement with those calculated for pH = 7 by the SPARC program (3.7); this program, however, predicts higher log K_{oc} with decreasing pH. With decreasing pH values, protonated species of MDA become more and more important. Based on the log K_{ow} alone, the program KOCWIN v2.1 program (US Environmental Protection Agency) delivers a value of

$$\log K_{oc} = 0.55315 \times \log K_{ow} + 0.9215 = 1.7392,$$

which is much lower than the measured values. Therefore, log K_{ow} alone is a poor predictor for log K_{oc}. The experimental apparent K_{oc} for MDA increases with time which can be allocated to the formation of non-extractable residues (see above). Therefore, reported K_{oc} values are of limited usability due to yet unknown interrelation between hydrophobic interaction, ion exchange, and chemical reaction. The formation of non-extractable residues, ion addition, is very poor in anaerobic, highly reduced soils. In case of specific questions, p. e. for the evaluation of a certain environmental contamination, specific measurements with soil species under question may be necessary.

In a short-term test with hydroponic cultures, MDA is rapidly adsorbed by tomato roots. The radioactivity of the MDA seems to stick to the surface of the roots. Whether MDA is further transported into plant tissue if exposure time is extended is not clear at the moment. Experience from several other aromatic amines indicates that MDA is expected to be rapidly metabolized in plant tissue to soluble and insoluble products. Hydroponic cultures with sorghum, bean, orchard grass, and carrots exposed to dissolved 4,4′-methylene-bis(2-chloroaniline) (MBOCA) showed strong adsorption to the roots but very limited transport to other parts of the plants (Voorman and Penner 1986b); bean and cucumber grown on soil contaminated with radiolabeled MBOCA showed hardly any transport of the radioactivity to aerial parts, but again strong adsorption onto the roots. 4-Chloroaniline, which has a higher electron density of the aromatic ring than 3,4-dichloroaniline, shows an increased binding to lignin structures (Harms and Langenbartels 1986). This is in concordance with observations of other investigators: the logarithm of the apparent first-order rate constant for the irreversible absorption is anti-proportional to the half-wave oxidation potential of the primary, aromatic amine (Li et al. 2000). Therefore, it is conceivable that some of the MDA taken up by plants is bonded to lignin structures in plants, forming insoluble residues. Overall, it is concluded that MDA in soil pore water is rapidly adsorbed by plant roots, but it seems to stick to the surface of the roots. Why this is the case cannot be answered at the moment. Experiments may have been too short to conclude that there is no transport in over-soil parts of the plant; however, this may be concluded by analogy to other primary aromatic amines.

Environmental modeling with the EUSES 2.1 program indicated that MDA achieves highest environmental concentrations in agricultural soil. MDA is not expected to be persistent under different environmental conditions; in highly reduced anaerobic soils, however, persistency of MDA or its metabolites cannot be ruled out. Bioaccumulation is unlikely for MDA, as indicated by its physical–chemical properties, BCF test results, and by metabolism tests with rats and monkeys. MDA is very toxic to aquatic life.

11 Summary

MDA is an aromatic amine and high production volume chemical. In this review, data on its environmental behavior have been summarized and evaluated.

Owing to the physical properties, MDA distributes into the water, sediment, and soil compartments when released into the environment, and due to the chemical reactivity it is chemically absorbed in oxygenated soils and sediments; in highly reduced soils and sediments, chemisorption is negligible (West et al. 2002).

The half-life time of MDA in air – the less relevant compartment – is about 6 to 7 h when the concentration of hydroxide radicals is about $10^6/cm^3$.

MDA was not readily degradable in most of the reported biodegradation screening tests. The good DOC removal in an OECD 302B test with sludge from an industrial wastewater treatment plant (BASF 1981), and tests with selective media and enrichment cultures (Kim et al. 2002) indicate that microorganisms can adapt to MDA as food source, which may explain the different findings in ready biodegradability tests, ranging from readily biodegradable to not inherent degradable. Also rates of CO_2 formation indicate that in general MDA can be mineralized in soil

(Cowen et al. 1996), but that side reactions are in competition with mineralization. Literature from other primary aromatic amines indicates that binding to organic matter in soil reduces the rate of biodegradation to a level typical for soil humic matter (Saxena and Bartha 1983).

In environmental water and sediment samples, the half-life for primary MDA decay is about 3–10 days (Schaefer and Ponizovsky 2013), but MDA is not completely mineralized. Metabolites formed are partly less polar, partly more polar than MDA itself (Schaefer and Ponizovsky 2013; West et al. 2002). 4,4′-Diaminobenzophenone and 4-aminobenzaldehyde have been reported as metabolites of MDA, but several other compounds are still not identified. Microbial activity influences the pattern of metabolite formation.

In water, MDA is not subject to hydrolysis (Ekici et al. 2001a), but may photochemically be transformed to 4,4′-diaminobenzophenone and 4,4′-diamino biphenyl (Ekici et al. 2001b).

Therefore, not only covalent binding to soil and sediment organic matter, but also oxidation reactions are in competition with biotic mineralization of MDA. Whereas the products of irreversible covalent binding to organic matter are not deemed to be of major concern, other products deserve a closer look. Decay of MDA is not necessarily a detoxification unless it is mineralization.

In a short-term test with hydroponic cultures, MDA is rapidly adsorbed by tomato roots. The radioactivity of the MDA seems to stick to the surface of the roots. Whether MDA is further transported into plant tissue if exposure time is extended is not clear at the moment. Experience from several other aromatic amines indicates that MDA is expected to rapidly being metabolized in plant tissue to soluble and insoluble products.

In terms of PBT criteria, MDA is toxic, but not bioaccumulating; in Japan, it would be regarded as persistent due to poor degradability in the relevant MITI tests. In the EU, DT_{50} values do not rate the substance persistent. However, not all of the major metabolites of MDA were identified, so the discussion of persistency is not finalized, yet.

Acknowledgement This work was sponsored by the International Isocyanate Institute, Inc. The views presented in this paper are those of the authors and not necessarily those of the sponsor.

Conflict of Interest T. Schupp worked for BASF, an MDA producer, until 2012.
 H. Allmendinger is a consultant for Currenta GmbH & Co. OHG.
 S. Shen is working for Dow, an MDA producer.
 B. T. A. Bossuyt is working for Huntsman, an MDA producer.
 C. Boegi and B. Hidding are working for BASF, an MDA producer.
 B. Tury and R. J. West have been employed by the International Isocyanate Institute, Inc.

Annex 1: Output of the QSAR Toolbox Microbial Metabolism Simulator for 4,4′-MDA

Smiles notation: Nc1ccc(Cc2ccc(O)cc2)cc1

Structure	Log K_{ow}	Vapor pressure (Pa/25°C)	Water solubility (mg/L)
	2.62	2.47E-04	975

Smiles notation: Nc1ccc(Cc2ccc(O)c(O)c2)cc1

Structure	Log K_{ow}	Vapor pressure (Pa/25°C)	Water solubility (mg/L)
	2.14	3.42E-06	2,070

Smiles notation: Nc1ccc(Cc2ccc(N)c(O)c2O)cc1

Structure	Log K_{ow}	Vapor pressure (Pa/25°C)	Water solubility (mg/L)
	1.22	1.16E-07	2,748

Smiles notation: Nc1ccc(Cc2ccc(O)cc2)c(O)c1O

Structure	Log K_{ow}	Vapor pressure (Pa/25°C)	Water solubility (mg/L)
	1.66	2.12E-07	4,372

Smiles notation: Oc1ccc(Cc2ccc(O)c(O)c2)cc1

Structure	Log K_{ow}	Vapor pressure (Pa/25°C)	Water solubility (mg/L)
	2.57	5.74E-06	866

Smiles notation: Oc1ccc(Cc2ccc(O)c(O)c2)cc1O

Structure	Log K_{ow}	Vapor pressure (Pa/25°C)	Water solubility (mg/L)
	2.09	3.89E-07	1,830

Smiles notation: Oc1ccc(Cc2ccc(O)c(O)c2O)cc1O

Structure	Log K_{ow}	Vapor pressure (Pa/25°C)	Water solubility (mg/L)
	2.04	2.52E-08	1,678

Smiles notation: OC(=O)C(O)=C(O)C=CC(=O)Cc1ccc(O)c(O)c1

Structure	Log K_{ow}	Vapor pressure (Pa/25°C)	Water solubility (mg/L)
	0.05	3.71E-12	>1.00E05

Smiles notation: OC=O

Structure	Log K_{ow}	Vapor pressure (Pa/25°C)	Water solubility (mg/L)
	−0.54	4.78E03	>1.00E05

Smiles notation: OC(CCC(O)=O)CC(=O)C(O)=O

Structure	Log K_{ow}	Vapor pressure (Pa/25°C)	Water solubility (mg/L)
	−2.64	4.94E-06	>1.00E05

Smiles notation: OC(C(O)=O)C(=O)C=CC(=O)Cc1ccc(O)c(O)c1

Structure	Log K_{ow}	Vapor pressure (Pa/25°C)	Water solubility (mg/L)
	0.27	5.85E-10	>1.00E05

Smiles notation: OC(=O)CCC=O

Structure	Log K_{ow}	Vapor pressure (Pa/25°C)	Water solubility (mg/L)
	−0.42	16	>1.00E05

Smiles notation: OC(=O)CCC(O)=O

Structure	Log K_{ow}	Vapor pressure (Pa/25°C)	Water solubility (mg/L)
	−0.59	2.55E-05	>1.00E05

Smiles notation: OC(=O)Cc1ccc(O)c(O)c1

Structure	Log K_{ow}	Vapor pressure (Pa/25°C)	Water solubility (mg/L)
	0.98	1.32E-04	>1.00E05

Smiles notation: OC(=O)CC=CCC(=O)C(O)=O

Structure	Log K_{ow}	Vapor pressure (Pa/25°C)	Water solubility (mg/L)
	−1.32	1.61E-03	>1.00E05

Smiles notation: OC(=O)CC=CC=C(O)C(O)=O

Structure	Log K_{ow}	Vapor pressure (Pa/25°C)	Water solubility (mg/L)
	−0.29	6.38E-06	1.36E05

Smiles notation: OC(=O)CC(C=O)=CC=C(O)C(O)=O

Structure	Log K_{ow}	Vapor pressure (Pa/25°C)	Water solubility (mg/L)
	−1.23	2.56E-07	>1.00E05

Smiles notation: Nc1ccc(Cc2ccc(O)c(O)c2)c(O)c1O

Structure	Log K_{ow}	Vapor pressure (Pa/25°C)	Water solubility (mg/L)
	1.18	1.37E-08	9,191

Smiles notation: CCC(O)=O

Structure	Log K_{ow}	Vapor pressure (Pa/25°C)	Water solubility (mg/L)
	0.33	471	>1.00E05

Smiles notation: CC(=O)C(O)=O

Structure	Log K_{ow}	Vapor pressure (Pa/25°C)	Water solubility (mg/L)
	−1.24	172	>1.00E05

Smiles notation: C=O

Structure	Log K_{ow}	Vapor pressure (Pa/25°C)	Water solubility (mg/L)
	0.35	5.19E05	>1.00E05

Smiles notation: C=CC=O

Structure	Log K_{ow}	Vapor pressure (Pa/25°C)	Water solubility (mg/L)
	−0.01	3.65E04	>1.00E05

Smiles notation: OC(=O)C=C

Structure	Log K_{ow}	Vapor pressure (Pa/25°C)	Water solubility (mg/L)
	0.35	529	53,468

Smiles notation: C=CC(=O)C=O

Structure	Log K_{ow}	Vapor pressure (Pa/25°C)	Water solubility (mg/L)
	−0.36	1,370	>1.00E05

Smiles notation: OC(C(O)=O)C(=O)C=C

Structure	Log K_{ow}	Vapor pressure (Pa/25°C)	Water solubility (mg/L)
	−0.65	3.86E-02	>1.00E05

Smiles notation: OC(=O)C(=O)C=C

Structure	Log K_{ow}	Vapor pressure (Pa/25°C)	Water solubility (mg/L)
	−0.11	32	>1.00E05

Annex 2: Summary Table of Selected EUSUS 2.1 Calculation Results

Section/parameter	Favorable	Unfavorable
Study		
Study identification		
Study name	MDA modeling favorable	MDA modeling unfavorable
Study description	MDA modeling	MDA modeling
Author	Thomas Schupp	Thomas Schupp
Institute	Muenster University of Applied Science	Muenster University of Applied Science
Address	Stegerwaldstrasse 39	Stegerwaldstrasse 39
Zip code	D-48565	D-48565
City	Steinfurt	Steinfurt
Country	Germany	Germany
Email	Thomas.schupp@fh-muenster.de	Thomas.schupp@fh-muenster.de
Calculations checksum	1E48A198	FC7A72C8
Substance		
Substance identification		
General name	Methylene-4,4′-dianiline	Methylene-4,4′-dianiline
Description		
CAS-No	101–77-9	101–77-9
Physico-chemical properties		
Molecular weight (g/mol)	198	198
Melting point (°C)	90	90
Boiling point (°C)	400	400
Vapor pressure at test temperature (Pa)	2.5E-04	2.5E-04
Temperature at which vapor pressure was measured (°C)	25	25
Vapor pressure at 25 (°C)	2.5E-04	2.5E-04
Octanol–water partition coefficient	1.55	1.55
Water solubility at test temperature (mg/L)	1,000	1,000
Temperature at which solubility was measured (°C)	25	25
Water solubility at 25 (°C)	1,000	1,000
Partition coefficients and bioconcentration factors		
Solids–water		
Chemical class for Koc-QSAR	Anilines	Anilines
Organic carbon–water partition coefficient (L/kg)	7.00E + 03	7.00E + 03

(continued)

Section/parameter	Favorable	Unfavorable
Degradation and transformation rates		
Charactarization		
Characterization of biodegradability	Readily biodegr., failing 10-d window	Inherently biodegr., fulfilling criteria
STP		
Rate constant for biodegradation in STP (h^{-1})	0.3	0.1
Water/sediment		
Water		
Rate constant for photolysis in surface water (day^{-1})	3.6E-04	3.6E-04
Rate constant for biodegradation in surface water $(day^{-1}, 12°C)$	0.0924	2.16E-03
Sediment		
Rate constant for biodegradation in aerated sediment $(day^{-1}, 12°C)$	0.224	2.84E-04
Air		
Rate constant for degradation in air (day^{-1})	0.7	0.7
Soil		
Rate constant for biodegradation in bulk soil $(day^{-1}, 20°C)$	5.11E-03	0
Release estimation		
Characterization and tonnage		
High Production Volume Chemical	Yes	Yes
Production volume of chemical in EU (t/a)	5.00E+05	5.00E+05
Use patterns		
Production steps		
Emission input data		
Emission scenario	No special scenario selected/available	No special scenario selected/available
Intermediate results		
Intermediate		
Release fractions and emission days		
Production		
Emission tables	A1.1 (general table), B1.6 (general table)	A1.1 (general table), B1.6 (general table)
Distribution		
Regional, continental, and global distribution		
Pecs		
Regional		
Regional PEC in surface water (total) (mg/L)	3.42E-03	0.0289
Regional PEC in seawater (total) (mg/L)	3.3E-04	2.79E-03
Regional PEC in surface water (dissolved) (mg/L)	3.38E-03	0.0286

(continued)

Section/parameter	Favorable	Unfavorable
Qualitative assessment might be needed (TGD Part II, 5.6)	No	No
Regional PEC in seawater (dissolved) (mg/L)	3.29E-04	2.79E-03
Qualitative assessment might be needed (TGD Part II, 5.6)	No	No
Regional PEC in air (total) (mg/m^3)	2.02E-09	2.04E-09
Regional PEC in agricultural soil (total) (mg/kg wet wt)	0.0557	7.1
Regional PEC in pore water of agricultural soils (mg/L)	4.51E-04	0.0574
Regional PEC in natural soil (total) (mg/kg wet wt)	1.42E-03	0.0428
Regional PEC in industrial soil (total) (mg/kg wet wt)	0.146	4.38
Regional PEC in sediment (total) (mg/kg wet wt)	0.111	7.98
Regional PEC in seawater sediment (total) (mg/kg wet wt)	4.48E-03	0.684

References

Alport DC, Gilbert DS, Outterside SM (eds) (2003) MDI and TDI: safety, health, and the environment – a source book and practical guide. Wiley, West Sussex

BASF (1981) Biologische Eliminierbarkeit im Zahn-Wellens-Test. BASF-SE, Experimental Toxicology and Ecology, Report No. 87/0892. Robust summary: http://apps.echa.europa.eu/registered/data/dossiers/DISS-9c7b34e3-c5c4-7414-e044-00144f67d249/AGGR-acfea30d-ad97-4274-b8ce-aabd18d556da_DISS-9c7b34e3-c5c4-7414-e044-00144f67d249.html

Baumann W (1985) Report on the test for ready biodegradability of TK10504 in the modified Sturm test (OECD Guideline No. 301B, Paris 1981). Ciba-Geigy Ltd, Basel, report R-1066.K.247; project no. 850729. 12.09.1985. Robust summary: http://apps.echa.europa.eu/registered/data/dossiers/DISS-9c7b34e3-c5c4-7414-e044-00144f67d249/AGGR-acfea30d-ad97-4274-b8ce-aabd18d556da_DISS-9c7b34e3-c5c4-7414-e044-00144f67d249.html

Baumann W (1986) Report on the bioelimination test on TK 10504 in the simulation test – aerobic sewage OECD Coupled Units Test No. 303 A. Ciba-Geigy Ltd, Basel. Project-no. 860630. Archive R-1066.K2.47. 15.10.1986. Robust summary: http://apps.echa.europa.eu/registered/data/dossiers/DISS-9c7b34e3-c5c4-7414-e044-00144f67d249/AGGR-acfea30d-ad97-4274-b8ce-aabd18d556da_DISS-9c7b34e3-c5c4-7414-e044-00144f67d249.html

Becker KH, Bastian V, Klein T (1988) The reaction of OH radicals with toluene diisocyanate, toluenediamine and methylenedianiline under simulated atmospheric conditions. J Photochem Photobiol A Chem 45:195–205

Bollag J-M, Minard RD, Liu S-Y (1983) Cross-linkage between anilines and phenolic humus constituents. Environ Sci Technol 17:72–80

Bollag J-M, Myers CJ, Minard RD (1992) Biological and chemical interactions of pesticides with soil organic matter. Sci Total Environ 123/124:205–217

Bongartz R (2012) MDA: uptake by plants in a hydroponic culture. III Report No. 11635. International Isocyanate Institute, Manchester. Robust summary: http://apps.echa.europa.eu/registered/data/dossiers/DISS-9c7b34e3-c5c4-7414-e044-00144f67d249/AGGR-acfea30d-ad97-4274-b8ce-aabd18d556da_DISS-9c7b34e3-c5c4-7414-e044-00144f67d249.html

Briggs GG, Olgivie SY (1971) Metabolism of 3-chloro-4-methoxyaniline and some N-acyl derivatives in soil. Pest Sci 2:165–168

Burge WD (1972) Microbial populations hydrolyzing propanil and accumulation of 3,4-dichloroaniline and 3,3′,4,4′-tetrachloro-azobenzene in soil. Soil Biol Biochem 4:379–386

Campbell K (2017) [14C]-methylenedianiline: generation of transformation products in aerobic aquatic sediment systems for identification by LC-MS/MS. Unpublished Report No. 11688 of the International Isocyanates Institute, Boonton

Carvajal-Diaz J (2015) IHS chemical economics handbook: aniline. IHS, Englewood

Caspers N, Hamburger B, Kanne R, Klebert W (1986) Ecotoxicity of toluene diisocyanate (TDI), diphenylmethane diisocyanate (MDI), toluene diamine (TDA), diphenylmethanediamine (MDA). III Report No. 10417. International Isocyanate Institute, Manchester. Available from: British Library Document Supply Centre, Boston Spa, Wetherby, West Yorks. Robust summary: http://apps.echa.europa.eu/registered/data/dossiers/DISS-9c7b34e3-c5c4-7414-e044-00144f67d249/AGGR-acfea30d-ad97-4274-b8ce-aabd18d556da_DISS-9c7b34e3-c5c4-7414-e044-00144f67d249.html

CITI (1992) 4-Methylphenylene-1,3-diamine (CAS-No. 95-80-7) & 4,4′-diaminodiphenylmethane (CAS-No. 101-77-8) [Acute toxicity, biodegradability and bioaccumulation data]. In: Biodegradation and bioaccumulation data of existing chemicals based on the CSCL Japan. Chemical Inspection and Testing Institute Japan, Japan Chemical Industry Ecology-Toxicology and Information Center, Tokyo, pp 3–24, 4–6. ISBN 4-89074-101-1

Cowen WF, Gastinger AM, Spanier CE, Buckel JR, Bailey RE (1996) Sorption and microbial degradation of toluene diamines and methylene dianiline in soil under aerobic and anaerobic conditions. III Report No. 11230. International Isocyanate Institute, Manchester. Robust summary: http://apps.echa.europa.eu/registered/data/dossiers/DISS-9c7b34e3-c5c4-7414-e044-00144f67d249/AGGR-acfea30d-ad97-4274-b8ce-aabd18d556da_DISS-9c7b34e3-c5c4-7414-e044-00144f67d249.html

Cowen WF, Gastinger AM, Spanier CE, Buckel JR (1998) Sorption and microbial degradation of toluenediamines and methylendianiline in soil under aerobic and anaerobic conditions. Environ Sci Technol 1998;32:598–603

Deng Y, Xu L, Sun X, Cheng L, Liu G (2015) Measurement and correlation of the solubility for 4,4′-diaminodiphenylmethane in different solvents. J Chem Eng Data 2015;60(7):2028–2034

ECHA (2014a) 4,4′-Methylenedianiline [REACh Dossier]. European Chemicals Agency, Helsinki. https://echa.europa.eu/de/registration-dossier/-/registered-dossier/15201. Accessed 21 June 2014

ECHA (2014b) Formaldehyde, oligomeric reaction products with aniline [REACH Dossier]. European Chemicals Agency, Helsinki. https://echa.europa.eu/de/registration-dossier/-/registered-dossier/14114. Accessed 21 June 2014

Ekici P, Leupold G, Parlar H (2001a) Degradability of selected azo dye metabolites in activated sludge systems. Chemosphere 44:721–728

Ekici P, Angerhoefer D, Parlar H (2001b) Photoinduced reactions of selected azo dye metabolites in water. Fresenius Environ Bull 10(3):245–256

European Union (2001) European Union risk assessment report, 4,4′-methylenedianiline, CAS-No. 101-77-9, EINECS-No. 202-974-4. Office for Official Publications of the European Union, Luxembourg. ISBN: 92-894-0484-1

European Union (2016) Joint Research Center: the European Union System for the Evaluation of Substances (EUSES). https://ec.europa.eu/jrc/en/scientific-tool/european-union-system-evaluation-substances. Accessed 29 Dec 2016

Frank R, Kloepffer W (1988) Spectral solar photon irradiation in Central Europa and the adjacent North Sea. Chemosphere 17(5):985–994

Government of Canada (2014) Draft screening assessment for methylenediphenyl diisocyanates and methylenediphenyl diamines, August 2014. Government of Canada, Chemicals Management Plan Division, Gatineau. http://www.chemicalsubstanceschimiques.gc.ca/group/diisocyanate-eng.php

Graveel JG, Sommers LE, Nelson DW (1985) Sites of benzidine, α-naphthylamine and p-toluidine retention in soil. Environ Toxicol Chem 4: 607–613

Han X, Nabb DL, Mingoia RT, Yang C-H (2007) Determination of xenobiotic intrinsic clearance in freshly isolated hepatocyte from rainbow trout (*Oncorhynchus mykiss*) and rat and its application in bioaccumulation risk assessment. Environ Sci Technol 41:3269–3276

Harms H, Langenbartels C (1986) Standardized plant cell suspension test systems for an ecotoxicologic evaluation of the metabolic fate of xenobiotics. Plant Sci 45: 157–165

Hellpointer E (1997) Determination of the quantum yield and assessment of the environmental half-life of the direct photodegradation of 4,4′-methylenedianiline in water. III Report No. 11265. International Isocyanate Institute, Manchester. Available from: British Library Document Supply Centre, Boston Spa, Wetherby, West Yorks. Robust study summary: http://apps.echa.europa.eu/registered/data/dossiers/DISS-9c7b34e3-c5c4-7414-e044-00144f67d249/AGGR-acfea30d-ad97-4274-b8ce-aabd18d556da_DISS-9c7b34e3-c5c4-7414-e044-00144f67d249.html

Howard PH, Boethling RS, Jarvis WF, Meylan WM, Michalenko EM (eds) (1991) Handbook of environmental degradation rates. Lewis Publishers, Chelsea

Hsu T-S, Bartha R (1974) Biodegradation of chloroaniline-humus complexes in soil and in culture solution. Soil Sci 118:213–220

Kajbaf M, Sepai O, Lamb JH (1992) Identification of metabolites of 4,4′-diaminodiphenylmethane (methylene dianiline) using liquid chromatographic and mass spectrometric techniques. J Chromat B 583:63–76

Karabunarliev S, Dimitrov S, Pavlov T, Nedelcheva D, Mekenyan O (2012) Simulation of chemical metabolism for fate and hazard assessment. IV. Computer-based derivation of metabolic simulators from documented metabolism maps. SAR QSAR Environ Res 23:371–387

Kaufman DD, Plimmer JR, Klingebiel UI (1973) Microbial oxidation of 4-chloroaniline. J Agric Food Chem 21(1):127–132

Kim M-N, Jang J-C, Lee I-M, Lee H-S, Yoon J-S (2002) Toxicity and biodegradation of diamines. J Environ Sci Health B 37(1):53–64

Li H, Lee LS, Jafvert CT, Graveel JG (2000) Effect of substitution on irreversible binding and transformation of aromatic amines with soils in aqueous systems. Environ Sci Technol 34:3674–3680

Li H, Lee LS, Schulze DG, Guest CA (2003) Role of manganese in the oxidation of aromatic amines. Environ Sci Technol 37:2686–2693

Mei C-F, Liu Y-Z, Long W-N, Sun G-P, Zeng G-Q, Xu M-Y, Luan T-G (2015) A comparative study of biodegradability of a carcinogenic aromatic amine (4,4′-diaminodiphenylmethane) with OECD 301 test methods. Ecotoxicol Environ Saf 111:123–130

Morgott DA (1984) The in vivo biotransformation and acute hepatotoxicity of methylene dianiline. Dissertation, University of Michigan

Nabb DL, Mingoia RT, Yang CH, Han X (2006) Comparison of basal level metabolic enzyme activities of freshly isolated hepatocytes from rainbow trout (*Oncorhynchuss mykiss*) and rat. Aquat Toxicol 80:52–9

NITE (Japan National Institute of Technology and Evaluation) (2007) 4,4-MDA risk assessment report. http://www.nite.go.jp/chem/chrip/chrip_search/dt/pdf/CI_02_001/risk/pdf_hyoukasyo/340riskdoc.pdf

NTP (2018) Methylenedianiline (101-77-9). Chemical Effects in Biological Systems (CEBS). Research Triangle Park, NC (USA): National Toxicology Program (NTP). https://manticore.niehs.nih.gov/cebssearch/test_article/101-77-9. Accessed 14 Mar 2018

Ononyne AI, Graveel JG (1994) Modelling the reaction of 1-napthylamine and 4-methylaniline with humic acids: spectroscopic investigations of the covalent linkages. Environ Toxicol Chem 13(4):537–541

Parris GE (1980) Covalent binding of aromatic amines to humates. 1. Reactions with carbonyls and quinones. Environ Sci Technol 14(9):1099–1106

Pemberton D, Tury B (2008) Prediction of atmospheric half-lives for MDI, TDI, MDA and TDA using the AOPWIN™ model. GIL Report No. 2008/C. Global Isocyanates, Manchester. Robust study summary: http://apps.echa.europa.eu/registered/data/dossiers/DISS-9c7b34e3-c5c4-7414-e044-00144f67d249/AGGR-acfea30d-ad97-4274-b8ce-aabd18d556da_DISS-9c7b34e3-c5c4-7414-e044-00144f67d249.html

Pillai P, Helling CS, Dragun J (1982) Soil-catalyzed oxidation of aniline. Chemosphere 11(3):299–317

Saxena A, Bartha R (1983) Microbial mineralization of humic acid – 3,4-dichloroaniline complexes. Soil Biol Biochem 15(1): 59–62

Schaefer EC, Carpenter K (2013) 4,4′-MDA: aerobic mineralization in surface waters. III Report No. 11652. International Isocyanate Institute, Manchester. Robust study summary: http://apps.echa.europa.eu/registered/data/dossiers/DISS-9c7b34e3-c5c4-7414-e044-00144f67d249/AGGR-acfea30d-ad97-4274-b8ce-aabd18d556da_DISS-9c7b34e3-c5c4-7414-e044-00144f67d249.html

Schaefer EC, Ponizovsky A (2013) 4,4′-MDA: aerobic and anaerobic transformation aquatic sediment systems. III Report No. 11646. International Isocyanate Institute, Manchester. Robust study summary: http://apps.echa.europa.eu/registered/data/dossiers/DISS-9c7b34e3-c5c4-7414-e044-00144f67d249/AGGR-acfea30d-ad97-4274-b8ce-aabd18d556da_DISS-9c7b34e3-c5c4-7414-e044-00144f67d249.html

Schupp T, Allmendinger H, Bossuyt BTA, Hidding B, Tury B, West RJ (2016) Review of the ecotoxicological properties of the methylenedianiline substances. Rev Environ Contam Toxicol 241: 39–72

Schwarz H (2009) MDA determination of the ready biodegradability according to OECD Test Guideline 301B. III Report No. 11567. International Isocyanate Institute, Manchester. Robust study summary: http://apps.echa.europa.eu/registered/data/dossiers/DISS-9c7b34e3-c5c4-7414-e044-00144f67d249/AGGR-acfea30d-ad97-4274-b8ce-aabd18d556da_DISS-9c7b34e3-c5c4-7414-e044-00144f67d249.html

Suidan MT, Campo P, Platten W, Chai Y, Davis JD (2011) Aerobic biodegradation of amines in industrial saline wastewaters. Chemosphere 85:1199–1203

US EPA (2004) Appendix C to 40 CFR Part 63 - determination of the fraction biodegraded (Fbio) in a biological treatment unit. Fed Regist 69:39383–39392

Voorman R, Penner D (1986a) Fate of MBOCA [4,4′-methylene-bis(2-chloroaniline)] in soil. Arch Environ Contam Toxicol 15: 595–602

Voorman R, Penner D (1986b) Plant uptake of MBOCA [4,4′-methylene-bis(2-chloroaniline)]. Arch Environ Contam Toxicol 15: 589–593

Weber EJ, Spidle DL, Thorn KA (1996) Covalent binding of aniline to humic substances. 1: kinetic studies. Environ Sci Technol 30:2755–2763

Weber EJ, Colon, D, Baughman, GL (2002) Sediment-associated reactions of aromatic amines. 2. QSAR development. Environ Sci Technol 36:2443–2450

West RJ (2007) Evaluation of the intrinsic anaerobic biodegradability of MDA and TDA in soil-free enrichment cultures. III Report No. 11530. International Isocyanate Institute, Manchester. Robust study summary: http://apps.echa.europa.eu/registered/data/dossiers/DISS-9c7b34e3-c5c4-7414-e044-00144f67d249/AGGR-acfea30d-ad97-4274-b8ce-aabd18d556da_DISS-9c7b34e3-c5c4-7414-e044-00144f67d249.html

West RJ, Tury B (2006) PBT categorization analysis for TDI, MDI, TDA and MDA. III Report No. 11522. International Isocyanate Institute, Manchester. Robust study summary: http://apps.echa.europa.eu/registered/data/dossiers/DISS-9c7b34e3-c5c4-7414-e044-00144f67d249/AGGR-acfea30d-ad97-4274-b8ce-aabd18d556da_DISS-9c7b34e3-c5c4-7414-e044-00144f67d249.html

West RJ, Davis JW, Bailey RE (2002) Sorption and biodegradation of 2,4-TDA, 2,6-TDA and 4,4′-MDA in soils under anaerobic conditions: second study. III Report No. 11463. International Isocyanate Institute, Manchester. Robust study summary: http://apps.echa.europa.eu/registered/data/dossiers/DISS-9c7b34e3-c5c4-7414-e044-00144f67d249/AGGR-acfea30d-ad97-4274-b8ce-aabd18d556da_DISS-9c7b34e3-c5c4-7414-e044-00144f67d249.html

Yakabe Y (1994) The study of the environmental fate of TDA, MDA and oligoureas of MDI and TDI: biodegradability test of 2,4-diaminotoluene and 4,4′-diaminodiphenylmethane. III Report No. 11170. International Isocyanate Institute, Manchester. Robust study summary: http://apps.echa.europa.eu/registered/data/dossiers/DISS-9c7b34e3-c5c4-7414-e044-00144f67d249/AGGR-acfea30d-ad97-4274-b8ce-aabd18d556da_DISS-9c7b34e3-c5c4-7414-e044-00144f67d249.html

Distribution of Microplastics and Nanoplastics in Aquatic Ecosystems and Their Impacts on Aquatic Organisms, with Emphasis on Microalgae

Jun-Kit Wan, Wan-Loy Chu, Yih-Yih Kok, and Choy-Sin Lee

Contents

Abbreviations

AChE   Acetylcholinesterase
Chl-a   Chlorophyll-a
CO_2 Carbon dioxide
DW   Dry weight
FTIR   Fourier transform infrared spectroscopy

J.-K. Wan (✉) · W.-L. Chu
School of Postgraduate Studies, International Medical University, Kuala Lumpur, Malaysia
e-mail: wanjunkit@gmail.com; wanloy_chu@imu.edu.my

Y.-Y. Kok
Applied Biomedical Science and Biotechnology Division, School of Health Sciences, International Medical University, Kuala Lumpur, Malaysia
e-mail: yihyih_kok@imu.edu.my

C.-S. Lee
Department of Pharmaceutical Chemistry, School of Pharmacy, International Medical University, Kuala Lumpur, Malaysia
e-mail: choysin_lee@imu.edu.my

© Springer International Publishing AG, part of Springer Nature 2018
P. de Voogt (ed.), *Reviews of Environmental Contamination and Toxicology Volume 246*, Reviews of Environmental Contamination and Toxicology 246, DOI 10.1007/398_2018_14

GPx   Glutathione peroxidase
GR   Glutathione reductase
GST   Glutathione S-transferases
HDPE   High-density polyethylene
MAPK   Mitogen-activated protein kinase
MP   Microplastics
NP   Nanoplastics
PC   Polycarbonate
PE   Polyethylene
PET   Poly(ethylene terephthalate)
PLA   Poly(lactic acid)
PMMA   Poly(methyl methacrylate)
PP   Polypropylene
PS   Polystyrene
PS-PEI   Polyethyleneimine polystyrene
PU   Polyurethane
PVC   Poly(vinyl chloride)
ROS   Reactive oxygen species
SOD   Superoxide dismutase
WW   Wet weight

1 Introduction

Plastics are widely used in transportation, packaging, construction and medical and health-care industries because of their diverse properties such as durability, easy and cheap to manufacture, resistant to shock, corrosion, chemicals and heat as well as being able to come in many different sizes, shapes and colours (da Costa et al. 2016; Lithner 2011). The top three plastic producers in the world are China (26%), Europe (20%) and North America (19%) (Crawford and Quinn 2017). Most of the plastics produced are polyethylene (PE) and polypropylene (PP) in which the packaging industry is the biggest consumer of these plastics (Crawford and Quinn 2017). Plastics have evolved with the addition of various other substances such as colorant, stabilizers, flame retardants and plasticizers (Lithner 2011; Talsness et al. 2009; Hammer et al. 2012). Examples of synthetic plastics include polystyrene (PS), PE, PP and poly(vinyl chloride) (PVC).

Indiscriminate and improper disposal of plastics lead to pollution in the aquatic environments. There have been reports on abandoned, lost and discarded fishing gears ("ghost gear"), tourism-related activities, effluents discharge as well as release of microfibres from conventional washing of garments, all of which cause plastics to be presented in the aquatic environments (Retama et al. 2016;

Phillips 2017; Hartline et al. 2016). Plastic pollution is a cause of great concern as some researchers reported the amount of plastics is even greater than the number of fish larvae in a major European river (Lechner et al. 2014). In 2016, the Ellen MacArthur Foundation reported in the World Economic Forum that by 2050, there will be more plastic than fish in the world's oceans, which was also reported by several newspapers worldwide (Kaplan 2016; Gosden 2016; Wearden 2016).

Plastics have been shown to be able to degrade to much smaller particles in micro- and nanosizes through various chemical and physical processes such as biodegradation, photodegradation, thermooxidative degradation and hydrolysis (Andrady 2011). These microscopic particles are termed "microplastics" (MP) (<5 mm) and "nanoplastics" (NP) (<1 μm) (da Costa et al. 2016). There are two categories of MP, namely, primary MP, which are a direct result of human material and product use, and secondary MP which are plastic fragments derived from the breakdown of larger plastic debris (Cole et al. 2011).

There are several methods to characterize and identify the type and composition of MP and NP in aquatic environments. Different types of samples such as sediments, sea and lake samples and raw and treated effluents had been collected from various study sites (Table 1). Soil and sediment samples are collected using methods such as quadrat sampling and bulk sampling. In quadrat sampling, the selected area of study is divided into quadrants, and the amounts of MP and NP collected are generally reported as abundance per unit of surface area, whereas in bulk sampling, samples are collected up to a specific sampling depth and amounts of MP and NP are generally reported as abundance per unit volume. Conversion between these two units of abundances is possible if the sampling depth is known; however, problems may arise because the sampling depth can be highly variable (Hidalgo-Ruz et al. 2012; Van Cauwenberghe et al. 2015). Floating MP and NP are generally collected using various nets such as neustonic and manta nets towed by research vessels (Miyake et al. 2011; Costa et al. 2010; Sadri and Thompson 2014; Morét-Ferguson et al. 2010). After collection, visual identification of plastic litter is carried out with the aid of a microscope in some studies, whereas others use density separation methods, Fourier transform infrared spectroscopy (FTIR), sequential pyrolysis gas chromatography coupled to mass spectrometry (pyrolysis GC/MS) and Raman spectroscopy (Tagg et al. 2015).

The primary aim of this paper is to review the distribution of MP and NP that have been detected in aquatic ecosystems, and the impacts of such pollutants on aquatic organisms. In addition, the impacts of contaminants such as heavy metals, persistent organic pollutants (POPs) and endocrine-disrupting chemicals (EDCs) associated with MP and NP in aquatic ecosystems are highlighted. Particular attention is also given to the potential impacts of MP and NP on microalgae since they form the basis of the food chain in aquatic ecosystems.

Table 1 Distribution of MP and NP in various aquatic ecosystems and wastewater discharge

Location	Type of sample(s)	Type of polymer(s)	Concentrations	References
Taihu Lake, China	Freshwater	Poly(ethylene terephthalate) (PET), polyester and PP	0.01×10^6–6.8×10^6 items/km^2 in plankton net samples 3.4–25.8 items/L in surface water 11.0–234.6 items/kg dry weight (DW) in sediments	Su et al. (2016)
River Meuse	Freshwater	PE, PS, PP	0.14 mg/m^3 or 9.7 MP/m^3	Urgert (2015)
River Rhine			0.56 mg/m^3 or 56 MP/m^3	
Wuhan, China	Freshwater	PET, PP, PE, polyamide and PS	$1,660 \pm 639$ to $8,925 \pm 1,591$ particles/m^3	Wang et al. (2017a)
Rhine river, Germany	Freshwater	PS, PP, polyester, PVC and poly (methyl methacrylate) (PMMA)	892,777 particles/km^2	Mani et al. (2014)
Deep Bay, Tolo Harbour, Tsing Yi, and Victoria Harbour, Hong Kong	Marine water	PP, PE, a blend of PP and ethylene propylene, and styrene acrylonitrile	51 to 27,909 particles/ 100 m^3	Tsang et al. (2017)
Bay of Calvi (Mediterranean–Corsica)	Marine water	PS	5.1 particles/100 m^2	Collignon et al. (2014)
Mobile Bay, Alabama, United States, Northern Gulf of Mexico estuaries	Marine water	PP, PE, PS, polyester and aliphatic polyamide	Seawater-dominated locations 50.60 ± 9.96 MP/m^2 Freshwater-dominated locations 13.20 ± 2.96 MP/m^2	Wessel et al. (2016)
Lagoon of Venice, Italy	Marine water	PE and PP	672 to 2,175 particles/ kg DW	Vianello et al. (2013)
Chesapeake Bay, United States	Marine water	PE and PS	<1.0 to >560 g/km^2	Yonkos et al. (2014)
Southern Ocean	Marine water	PE, PP, PE co-polymer, PS and PVC	3.1×10^{-2} pieces/m^3	Isobe et al. (2017)
Guangdong, South China	Marine water	PS	248–17,505 items/m^2 or 0.2730–16.9980 g/ m^2	Fok et al. (2017)
Deep Bay, Tolo Harbour, Tsing Yi, and Victoria Harbour, Hong Kong	Sediments	PP, PE, a blend of PP and ethylene propylene, and styrene acrylonitrile	49 to 279 particles/kg	Tsang et al. (2017)
Vembanad Lake, Kerala, India	Sediments	Low-density polyethylene (LDPE)	96–496 particles/m^2	Sruthy and Ramasamy (2017)
Beijiang River littoral zone, China	Sediments	PE, PP and copolymer	178 ± 69–544 ± 107 items/kg sediment	Wang et al. (2017b)
Sinop Sarikum Lagoon, Southern Black Sea	Sediments	PS	MP 1 mm to 5 mm 0.005–0.024 pieces/g MP < 1 mm 0.027–0.049 pieces/g	Visne and Bat (2016)
Deep Bay, Tolo Harbour, Tsing Yi, and Victoria Harbour, Hong Kong	Sediments	PP, PE, a blend of PP and ethylene propylene, and styrene acrylonitrile	49–279 particles/kg	Tsang et al. (2017)
Baltic Sea	Sediments	Polyester, poly(vinyl acetate) and poly(ethylene-propylene)	Beach sediment 25–53 particles/kg DW Bottom sediment 0–27 particles/kg DW	Graca et al. (2017)

(continued)

Table 1 (continued)

Location	Type of sample(s)	Type of polymer(s)	Concentrations	References
River Thames, UK	Sediments	PP, polyester and polyarylsulfone	66 particles/100 g	Horton et al. (2017a)
Cartagena, Colombia	Sediments	PE and PP	Total of 45,520 pellets	Acosta-Coley and Olivero-Verbel (2015)
River Scheldt, France, Belgium, and Netherlands	Sediments	PP, PE and PVC	$1,840 \pm 2,407$ to $63,112 \pm 24,628$ MP/kg DW	De Troyer (2015)
Rhine-Main area, Germany	Sediments	PE, PP and PS	1 g/kg or 4,000 particles/kg	Klein et al. (2015)
Kaliningrad region, Russia	Sediments	Foamed plastic	1.3–36.3 items/kg DW	Esiukova (2017)
St. Lawrence River, Canada	Sediments	PE	Mean $13,832 \pm 13,677$ microbeads/m^2	Castañeda et al. (2014)
Chicago metropolitan area of northeastern Illinois and northwestern Indiana and Central Illinois, United States	Wastewater treatment plant effluent	PP, PE and PS	1,338,757 pieces/day	McCormick et al. (2016)
Ireland	Wastewater treatment plant sludge	High-density polyethylene (HDPE), PE, polyester, acrylic, PET, PP and polyamide	4,196–15,385 particles/kg DW	Mahon et al. (2016)

2 Distribution of Microplastics and Nanoplastics in Aquatic Ecosystems

Various studies have been conducted to investigate the amounts of MP and NP present in aquatic environments and wastewater discharge (Table 1). In this review, only studies related to freshwater and marine ecosystems which fall within the operational definition of MP and NP as well as type of polymers (e.g. PS, PE, and PVC) stated are included. Studies in which there is overlap in size ranges of MP and mesoplastics and macroplastics are excluded.

As summarized in Table 1, the three most common types of MP and NP found are PP (16 study sites), followed by PE (15 study sites) and PS (11 study sites). Samples taken from the environment include freshwater, marine water, wastewater (treated and untreated) as well as sediments. Samples were taken from a diverse range of environments such as from heavily populated urban areas to sparsely populated regions. Microplastics and NP could even be found in very remote environments such as in mountainous lake and polar regions (Free et al. 2014; Waller et al. 2017; Lusher et al. 2015; Cincinelli et al. 2017). This demonstrates that sea and wind currents could impact on the distribution of MP and NP. The absence of proper waste disposal systems in remote regions could also contribute to MP and NP pollution.

It is difficult to compare the concentrations of MP and NP between different study sites due to different sampling procedures used and, hence, different abundance units. Even studies collecting similar type of samples (e.g. sediments) also reported abundances in different units. A standardized unit of measurement should be developed to facilitate easier comparison between various reported values. Some studies compiled data for a wide range of sizes of MP, with some more than 20 mm in diameter reported as MP (Waller et al. 2017; Haseler et al. 2017; Young and Elliott 2016).

According to da Costa et al. (2016), MP are microscopic particles <5 mm, while NP are <1 μm. However, in earlier studies, various size ranges had been reported as MP. For instance, Graham and Thompson (2009) defined MP as particles <10 mm, while Browne et al. (2007, 2010) and Claessens et al. (2011) reported MP as <1 mm. With this in mind, it is important for the size ranges of MP to be universally defined to facilitate easier comparison between various studies. Only two studies compared the differences between abundances of MP and NP in sediments and water column. Su et al. (2016) reported the abundances of MP reached 0.01×10^6–6.8×10^6 items/ km^2 in plankton net samples, 3.4–25.8 items/L in surface water, 11.0–234.6 items/kg dry DW in sediments and 0.2–12.5 items/g wet weight (WW) in Asian clams (*Corbicula fluminea*). There was a negative correlation ($p < 0.05$) between the contents of MP in clams and MP in the sediments, which indicated that clams accumulated MP even when MP concentration in the water column was low due to their extensive filter-feeding activities. In another study, Tsang et al. (2017) reported that there was more MP in coastal waters of Hong Kong (51–27,909 particles/100 m^3) compared to that in sediments (49–279 particles/kg).

3 Effects of Microplastics and Nanoplastics on Aquatic Organisms

There have been many reports in the literature on various laboratory and field studies assessing the effects of MP and NP on a wide range of organisms, including invertebrates (Table 2) and microalgae (Sect. 3.1; Table 3). The information derived from Table 2 shows that the most common synthetic polymer tested was PS (19 studies) followed by PET (5 studies). The test sizes of MP and NP ranged from as small as 20 nm to as large as 1,400 μm. Most of the studies focused only on the effects of a single size of MP or NP on the test organism. Particle size may influence the toxicity of MP and NP. For instance, Lee et al. (2013) tested the effects of PS beads over two generations of the copepod *Tigriopus japonicus* and found that PS beads of 0.05 μm were more toxic compared to that of 0.5 μm in terms of mortality rate, whereas 6 μm did not cause any mortality amongst copepods. Therefore, it is important to include a wide range of particle sizes in assessing the toxicity of MP and NP. In some studies, the size of MP and NP is too large to be ingested by the test organism, leading to no reported toxic effects (Rehse et al. 2016).

As summarized in Table 2, there were very few studies that reported the EC_{50}/ IC_{50} values of MP and NP, and none was based on minimum inhibition concentration (MIC). In most studies, EC_{50}/IC_{50} values could not be determined due to very

Table 2 Effects of microplastics and nanoplastics on aquatic organisms other than microalgae

Organisms	Type of polymer(s)	Size of particle	Concentrations	EC_{50}/IC_{50}	Effects	Reference
Freshwater crustacean (*Daphnia magna*)	PS	20, 1,000 nm	2 µg/L (24 h exposure)	N/A	Presence of MP in the gastrointestinal tract and translocation of MP to oil storage droplets	Rosenkranz et al. (2009)
	PET	62–1,400 µm	12.5–100 mg/L (48 h exposure and subsequent 24 h recovery)	N/A	Increased mortality of daphnids and failure of daphnids to recover from MP exposure	Jemec et al. (2016)
	PE	1, 100 µm	12.5–400 mg/L (96 h exposure)	57.43 mg/L (1 µm)	Ingestion of MP led to dose- and time-dependent immobilization	Rehse et al. (2016)
	PS	~70 nm	0.22–150 mg/L (21-day exposure)	27%	Reduced body size and severe alterations in reproduction, decreased numbers and body size of neonates, increased in number of neonate malformations	Besseling et al. (2014)
Freshwater crustacean (*Daphnia magna*), amphipod crustacean (*Corophium volutator*) and bacteria (*Vibrio fischeri*)	Pure PMMA and poly(-methylmethacrylate)-co-stearylmethacrylate (PMMA-PSMA) copolymer	86–125 nm	0.01, 0.1 and 1.0 mg/L and a high concentration 500–1,000 mg/L (48 h exposure)	879 mg/L (PMMA-PSMA plastic NP) 887 mg/L (fluorescent analogue of plastic NP)	Significant uptake and rapid excretion of the fluorescent NP was observed	Booth et al. (2015)
Zooplankton species (*Acartia tonsa*, Porcellanid larvae, *Calanus helgolandicus* and *Oithona similis*)	PS	20 µm	100 particles/mL (24 h exposure)	N/A	Energy deficit in all species with the exception of *O. similis*	Dedman (2014)
	PE	8.77–127.34 µm				
	Polyamide-6 nylon	8.83–123.42 µm				
	Nylon	8.58–134.56 µm				
Thirteen zooplankton taxa, including holoplankton, meroplankton and microzooplankton	PS	0.4–30.6 µm	1×10^6–635 beads/mL (1–24 h exposure)	N/A	Adherence to the external surfaces and decreased feeding rates	Cole et al. (2013)
Marine copepod (*Paracyclopina nana*)	PS	0.05, 0.5 and 6 µm	0–20 µg/mL (24 h exposure followed by 24 h recovery)	N/A	Developmental delays and decrease in fecundity, delayed moulting, increased ROS levels, phosphorylation of MAPK, and antioxidant enzymatic activities of GPx, GR, GST and SOD	Jeong et al. (2017)

(continued)

Table 2 (continued)

Organisms	Type of polymer(s)	Size of particle	Concentrations	EC$_{50}$/IC$_{50}$	Effects	Reference
Monogonont rotifer (*Brachionus koreanus*)	PS	0.05, 0.5 and 6 μm	0.1, 1, 10 and 20 μg/mL (12-day exposure) 10 μg/mL (48 h exposure)	N/A	Reduced growth rate, fecundity and lifespan and longer reproduction time. Antioxidant-related enzymes and MAPK signaling pathways were significantly activated	Jeong et al. (2016)
Marine copepod (*Calanus helgolandicus*)	PS	20 μm	75 MP/mL (24 h exposure)	N/A	Altered the feeding capacity and significantly reduced reproductive output	Cole et al. (2015)
Marine copepod (*Tigriopus japonicus*)	PS	0.05 μm	9.1×10^{11} particles/mL (96 h exposure) 2.1×10^{5} particles/mL (14-day exposure)	N/A	Mortality and significant reduction in fecundity	Lee et al. (2013)
		0.5 μm	9.1×10^{8} particles/mL (96 h exposure) 2.1×10^{5} particles/mL (14-day exposure)			
		6 μm	5.25×10^{5} particles/mL (96 h exposure) 2.1×10^{5} particles/mL (14-day exposure)			
Freshwater amphipod (*Hyalella azteca*)	PP	N/A	0–90 MP/mL (10-day exposure) 0–22.5 MP/mL (42-day exposure)	N/A	Reduced growth and reproduction	Talley (2015)
Beachhopper (*Platorchestia smithi*)	PE	38–45 μm	3.8% DW (72, 120 h exposure)	N/A	Significant effect on jump height and an increase in time to relocate shelter following disturbance	Tosetto (2015)
Lugworms (*Arenicola marina* (Linnaeus 1758))	Polylactic acid (PLA)	1.4–707 μm	0.02, 0.2 and 2% of sediment WW (31-day exposure)	N/A	Metabolic rates increased, while microalgal biomass decreased	Green et al. (2016)
	PE	2.5–316 μm				
	PVC	8.7–478 μm				

Organism	Polymer	Size	Concentration/exposure		Effects	Reference
Sea urchin embryo (*Paracentrotus lividus*)	PS	40 nm	0.02, 0.2 and 2% of sediment WW (31-day exposure) 2.5–50 µg/mL (48 h exposure)	N/A	PS-NH$_2$ caused severe developmental defects. *Abcb1* gene was upregulated at 48 h post fertilization (hpf) by PS-COOH, whereas PS-NH$_2$ induced *cas8* gene at 24 hpf	Della Torre et al. (2014)
		50 nm	1–50 µg/mL (48 h exposure)	3.82 µg/mL (24 hpf) 2.61 µg/mL (48 hpf)		
Barnacles (*Megabalanus azoricus*)	PVC	1.5 µm	0, 0.003, 0.03, 0.3 and 3% (6-week exposure)	N/A	Reduced cirral activity and oxygen consumption	Hentschel (2015)
Brine shrimp larvae (*Artemia franciscana*)	PS	40 nm 50 nm	5–100 µg/mL (48 h exposure, 24 h recovery)	N/A	Impaired food intake and motility and increased moulting events. Retention of MP up to 7 days	Bergami et al. (2016)
Zebrafish larvae (*Danio rerio*)	PS	45 µm 50 nm	1 mg/L (120 h exposure)	N/A	MP upregulated *zfrho* visual gene expression, whereas NP inhibited the larval locomotion and acetylcholinesterase activity and significantly reduced larvae body length and upregulated *gfap, α1-tubulin, zfrho* and *zfblue* gene expression significantly	Chen et al. (2017)
Blue mussel (*Mytilus edulis*)	PS	30 nm	0, 0.1, 0.2 and 0.3 g/L (8 h exposure)	N/A	Increased total weight of the feces and pseudofeces. Reduced filtering activity	Wegner et al. (2012)
Blue mussel (*Mytilus edulis*)	PS	2, 4–16 µm 3, 9.6 µm	0.51 g/L (12 h exposure) 15,000 particles/treatment	N/A	Accumulation of MP in their gut cavity and digestive tubules. Significant time-dependent effects in biological assays	Browne et al. (2008)
Blue mussel (*Mytilus edulis*)	HDPE	0–80 µm	2.5 g/L (96 h exposure)	N/A	Notable histological changes upon uptake and strong inflammatory response	Von Moos et al. (2012)
Pacific oyster (*Crassostrea gigas*)	PS	2, 6 µm	0.023 mg/L (2-month exposure)	N/A	Significant decrease in oocyte number, diameter and sperm velocity. Decrease in D-larval yield and larval development of offspring derived from exposed parents	Sussarellu et al. (2016)
European flat oysters (*Ostrea edulis*)	PLA PE	0.6–363 µm 0.48–316 µm	0.8, 80 µg/L (60-day exposure)	N/A	Benthic assemblage structures differed. Species richness, total number of organisms and biomass decreased	Green (2016)
Shore crabs (*Carcinus maenas*)	PS	8 µm	10^6, 10^7 microspheres/L (24 h exposure)	N/A	Presence of MP in gills, significant drop in the concentration of Na$^+$ ions within the haemolymph, and significant increase in Ca^{2+} and haemocyanin	Watts et al. (2016)

(continued)

Table 2 (continued)

Organisms	Type of polymer(s)	Size of particle	Concentrations	EC_{50}/IC_{50}	Effects	Reference
Crucian carp (*Carassius carassius*)	PS	24, 27 nm	1×10^{13} particles (130 mg particles per feeding) (61-day exposure)	N/A	Severe effects on feeding and shoaling behaviour and metabolism of the fish	Mattsson et al. (2014)
Three-spined stickleback (*Gasterosteus aculeatus*)	PS	1.0 μm	1.81×10^{10} particles/mL (7-day exposure, 14-day recovery)	N/A	Prolonged food digestion, effects on length, weight and condition index K were found	Katzenberger (2015)
		9.9 μm	1.81×10^{7} particles/mL (7-day exposure, 14-day recovery)			
		0.5 μm	1.4×10^{14} particles/mL (7-day exposure, 14-day recovery)			
Zebrafish (*Danio rerio*)	PS	5 μm	0–2,000 μg/L (3-week exposure)	N/A	Accumulation in fish, inflammation and lipid accumulation in liver. Induced significantly activities of SOD and CAT, alterations of metabolic profiles in fish liver and disturbed the lipid and energy metabolism	Lu et al. (2016)
		70 nm				
Kidney cells from fathead minnow (*Pimephales promelas*)	PS	41.0 nm	0.025, 0.05, 0.1, 0.2 μg/mL (2 h exposure)	N/A	Exposure of neutrophils caused significant increase in degranulation of primary granules and neutrophil extracellular traps (NETs) release	Greven et al. (2016)
	Polycarbonate (PC)	158.7 nm				

N/A not available

Table 3 Effects of microplastics and nanoplastics on microalgae

Microalgae	Type of polymer	Size of particle	Concentrations tested	EC_{50}/IC_{50} Value	Parameters measured	Effects
Microcystis aeruginosa and *Dolichospermum flos-aquae*	N/A	<200 μm	66.7 mg/L	N/A	Abundance, biovolume and colony size	No sustained effect on algal biomass and growth. Algal particle size was smaller at one point during the growth cycle
Skeletonema costatum	PVC	1 μm 1 mm	0–50 mg/L	N/A	Chlorophyll content and photosynthetic efficiency	Obvious inhibition on growth and the maximum growth inhibition ratio (IR) reached up to 39.7% after 96 h exposure. High concentration of MP (50 mg/L) had adverse effects on algal photosynthesis
Dunaliella tertiolecta	PS	0.05 μm 0.5 μm 6 μm	25, 250 mg/L	N/A	Photosynthetic efficiency and growth	Growth negatively affected up to 45% by uncharged particles but only at high concentration of 250 mg/L. Negligible effects on the photosynthetic efficiency
Pseudokirchneriella subcapitata	PS	110 nm	0–100 mg/L	>100 mg/L (72 h)	Growth	Negatively charged PS-COOH particles (110 nm) at 10 mg/L induced growth inhibition
	PS-PEI	55 nm 110 nm	0–100 μg/mL	0.58 μg/L (72 h) 0.54 μg/L (72 h)	Specific growth rate	Reduced specific growth rate. Particle of 110 nm was more toxic compared to particle of 55 nm
Scenedesmus obliquus	PS	~70 nm	44–1,100 mg/L	N/A	Photosynthetic capacity and biomass	Reduced chlorophyll concentrations
Tetraselmis chui	PS	1–5 μm	0.02–0.64 mg/L	0.145 mg/L (Cu + MP) (72 h)	Specific growth rate	No significant effects of MP. No significant differences between the toxicity curves of copper in the presence and absence of MP
Chlorella and *Scenedesmus*	PS	20 nm	0.08–0.8 mg/mL	N/A	Cellular respiration ROS production	Inhibited algal photosynthesis. ROS assay indicated that plastic adsorption promoted algal ROS production

N/A not available

low toxicity of MP and NP to the test organisms. Only studies involving planktonic organisms and invertebrates reported EC_{50}/IC_{50} values (Rehse et al. 2016; Booth et al. 2015). However, in other organisms such as shellfish and fish, despite low or no mortality, various biochemical and behavioural changes were also documented (Watts et al. 2016; Mattsson et al. 2014; Katzenberger 2015; Lu et al. 2016; Greven et al. 2016). Therefore, the toxicity of MP and NP may not follow the dose-response effect compared to other pollutants. Jemec et al. (2016) reported in their study to investigate the effect of MP textile fibres on *Daphnia magna* that the mortality of the crustacean did not increase with increasing concentrations of MP and was generally below 50%; thus, they were unable to calculate the LC_{50} values.

Despite the many reports in the literature, it is not possible to compare the effects of MP and NP across studies as the concentrations of toxicants are often expressed in different units, in mass/volume, mass/mass, particles/volume, beads/volume or MP/volume. Furthermore, the concentrations of MP and NP tested are beyond the levels that are environmentally relevant, sometimes higher by several orders of magnitude. For instance, Besseling et al. (2014) reported the nano-PS concentration thresholds inducing malformed *Daphnia magna* offsprings in their study to be a factor 10^6 higher than the 0.04–34 ng/L MP concentrations found in freshwater in Europe and the USA, and a factor 100 higher than the highest reported MP concentration in marine water.

Plastic debris has also been documented to be ingested by a wide range of organisms from plankton to whales (Table 2). In general, ingestion of MP and NP appears to be the principal entry mechanism of these particles into organisms such as zooplankton, invertebrates, shellfish and fish. This leads to reduced growth as well as various adverse effects on reproduction, feeding rates and malformation and eventual death. The transfer of MP and NP along the trophic levels of food chain is well documented. For instance, the seaweed *Fucus vesiculosus* was found to retain suspended MP on its surface, which were subsequently ingested by the common periwinkle *Littorina littorea,* as evidenced by the presence of MP in the stomach and gut and later excreted together with feces (Gutow et al. 2015).

Chae et al. (2018) investigated the trophic transfer of NP along four-species freshwater food chain. Nanoplastics were found adhered to the surface of the primary producer *Chlamydomonas reinhardtii,* which was subsequently ingested by planktonic *Daphnia magna* through filter feeding. The zooplankton was in turn predated upon by the secondary-consumer fish *Oryzias sinensis* and then consumed by the end-consumer fish *Zacco temminckii.* At every trophic level, NP were present in the organisms as confirmed by optical microscopy as well as confocal laser scanning microscopy. As *Chlamydomonas reinhardtii* is a motile alga, it would be interesting to assess the effect if similar experiment was conducted on nonmotile algae such as *Chlorella* sp. and *Pseudokirchneriella subcapitata.*

The reported studies on the transfer of MP and NP along the trophic levels (Gutow et al. 2015; Chae et al. 2018) may be too simplistic in reflecting the natural occurrence in the environment. Several food chains may interact to form a food web, and this adds to the complexity of the relationships between organisms at different trophic levels. Organisms at each trophic level in a food chain may not occupy the same level in another food chain, affecting the uptake of MP and NP. A mesocosm experimental design would be an appropriate approach to investigate the complex interactions between various food chains in relation to the transfer of MP and NP between organisms. For instance, Setälä et al. (2016) used the mesocosm approach to investigate the influence of feeding type on MP ingestion amongst invertebrates in a coastal community. The study found that bivalves contained significantly higher amounts of MP compared with free-swimming crustaceans and benthic, deposit-feeding animals. Besides that, free-swimming crustaceans ingested more MP compared with the benthic animals that fed only on the sediment surface.

3.1 Effects of Microplastics and Nanoplastics on Microalgae

Despite the abundant information on the effects of MP and NP on various organisms in the literature, there have been only a few studies that assessed the effects of MP and NP on microalgae (Table 3). The impact of MP and NP on microalgae is of pressing concern because as they are the primary producers, any potential toxic effects on them may affect other organisms at the higher trophic levels. The effects of MP and NP varied with the type of particles and algal species tested.

Long et al. (2017) investigated the interaction between virgin 2 μm PS beads and three microalgae, namely, *Tisochrysis lutea* (prymnesiophyte), *Heterocapsa triquetra* (dinoflagellate) and *Chaetoceros neogracile* (diatom). The authors found that the phytoplankton cells form hetero-aggregates with the micro-PS particles, depending on microalgal species and physiological status. Hetero-aggregation of the MP with *Chaetoceros neogracile* during the stationary growth phase was observed. No effects of micro-PS were observed on microalgal physiology in terms of growth and chlorophyll fluorescence. Hetero-aggregation of microalgae with MP may increase the particle density, leading to increased sedimentation rate of the particles. The MP could potentially be ingested by other organisms and transferred through the food web (Lagarde et al. 2016).

In another study, Yokota et al. (2017) found that the filamentous microalga *Dolichospermum flos-aquae* often formed bundles attached to corners of irregularly shaped MP, while the unicellular cyanobacterium *Microcystis aeruginosa* did not show such adhesion pattern. The study indicated that MP could potentially provide a suitable surface for colonization and/or maintenance of microalgae. Further, MP treatment did not affect the microalgal biomass and growth, although the algal particles were smaller in size at one point during the growth cycle, which did not persist throughout the 21-day experimental period.

A study to investigate the interaction between PVC (average diameter 1 μm) and *Skeletonema costatum* showed that the MP inhibited the growth of the diatom, with the highest growth inhibition ratio (IR) reaching up to 39.7% after 96 h exposure (Zhang et al. 2017). High concentration of MP (50 mg/L) also had adverse effects on algal photosynthesis as indicated by the decrease in chlorophyll content and photosynthetic efficiency (ɸ PSII). Shading effect was demonstrated not to be a reason for the toxicity of MP in this study. The authors postulated that physical damage due to precipitates resulting from the hetero-aggregation of MP and algal cells could be the contributing factor to the toxic effects observed. Adsorption of micro-PVC to the surface of cells may limit the transfer of energy and substances between cells and environment, leading to decrease in the uptake of nutrients and availability of light, CO_2 and O_2. Besides that, toxic metabolites may be retained inside the cell, causing the adverse effects.

Sjollema et al. (2016) reported the growth of three microalgae, namely, *Dunaliella tertiolecta*, *Thalassiosira pseudonana* and *Chlorella vulgaris*, was negatively affected (up to 45%) by uncharged PS particles but only at high concentration (250 mg/L). Besides that, these adverse effects were reported to increase with decreasing particle size. Both positively and negatively charged PS particles had negligible effects on photosynthetic efficiency of the three microalgae tested. However, a clear inhibitory effect of the uncharged PS beads on the growth of *Dunaliella tertiolecta* was observed. At the highest PS concentration of 250 mg/L, the average cell density of the *Dunaliella tertiolecta* exposed to 0.05 μm PS beads was clearly reduced (45%) compared to those exposed to 0.5 μm PS beads (11%) and 6 μm PS beads (<10%).

Nolte et al. (2017) reported that negatively charged PS-COOH particles (110 nm) at 10 mg/L inhibited growth of *Pseudokirchneriella subcapitata* with 72 h EC_{50} value exceeding 100 mg/L. They postulated that growth inhibition was caused by several possible factors such as nutrient depletion, increased osmotic pressure as well as immobilization of the algae due to the agglomeration of the cell culture. Casado et al. (2013) compared the inhibitory effect of polyethyleneimine polystyrene (PS-PEI) of two sizes (55 and 110 nm) on *Pseudokirchneriella subcapitata*. The larger PS-PEI particle (110 nm) induced greater growth inhibition compared to 55 nm. The authors suggested that the lower toxicity of 55 nm particles could be due to the higher degree of surface functionalization than the 110 nm NP. In another study, Besseling et al. (2014) reported that the growth inhibitory effect of nano-PS on the freshwater green alga *Scenedesmus obliquus* increased with increasing concentrations of the NP. Further, Davarpanah and Guilhermino (2015) noted that there were no significant effects of PS MP (0.046 to 1.472 mg/L) on the growth of the marine microalga *Tetraselmis chui*. The authors also assessed the combined effects of MP and copper and concluded that MP did not significantly affect the toxicity of the metal.

Physical adsorption of positively and negatively charged nanosized plastic beads onto two freshwater microalgae, *Chlorella* and *Scenedesmus*, was shown to inhibit photosynthesis, possibly through the physical blockage of light and airflow by the nanoparticles using a CO_2 depletion assay (Bhattacharya et al. 2010). The algal cells were incubated on a shaker separately with both positively and negatively charged PS beads of 0.08–0.8 mg/mL for 2 h at room temperature. Reactive oxygen species (ROS) assay further indicated that plastic adsorption promoted algal ROS production.

Most of the studies investigating the toxicity of MP and NP on microalgae focused on using PS. More studies needed to be carried out on other types of plastic polymers. Besides that, it is important to test the effects of several types of polymer on a single microalga as this is more representative of the actual environmental interaction between microalgae and MP and NP. Further, studies investigating the effects of MP and NP on microalgae reported contradictory results to each other. In some studies, uncharged MP and NP have no effects on microalgae (Long et al. 2017), but other studies reported adverse effects of such MP and NP (Sjollema et al. 2016).

In terms of microalgal photosynthesis, Sjollema et al. (2016) documented that PS beads do not affect microalgal photosynthesis even after 72 h. However, Bhattacharya et al. (2010) reported that both positively and negatively charged PS beads reduced photosynthesis in *Chlorella* and *Scenedesmus* after 4 h incubation with PS beads. This was supported by Besseling et al. (2014) and Zhang et al. (2017) who also reported that exposure to nano-PS beads for 72 h and micro-PVC for 96 h induced a decrease in chl-a content in *Scenedesmus obliquus*, and decrease in both chlorophyll content and photosynthetic efficiency in *Skeletonema costatum*. In addition, algal biomass and/or growth was reduced in *Thalassiosira pseudonana* and *Dunaliella tertiolecta* (Sjollema et al. 2016), *Pseudokirchneriella subcapitata* (Casado et al. 2013) and *Scenedesmus obliquus* (Besseling et al. 2014) but remained unaffected in *Tetraselmis chui* (Davarpanah and Guilhermino 2015) after short-term exposure to MP and NP. However, Yokota et al. (2017) reported that MP treatment had no sustained effect on algal biomass and growth of *Microcystis aeruginosa* and *Dolichospermum flos-aquae* even after 21-day exposure. Further studies to compare the effects of the duration of exposure to MP and NP on microalgae are worthwhile to provide a better understanding of this aspect.

4   Co-toxicity of Microplastics and Nanoplastics with Other Pollutants

Microplastics and NP are capable of concentrating various pollutants such as hydrophobic persistent organic pollutants (POPs) and heavy metals. Due to the large surface area to volume ratio and chemical composition of MP and NP, the concentrations of such chemicals can be up to the order of 10^6 greater than those present in seawater (Mato et al. 2001; Wagner et al. 2014). Besides that, various plastic additives originating from MP and NP could potentially leach out from the surface over time because these chemicals are weakly bound, or not bound at all to the polymer molecule (Horton et al. 2017b). Thus, MP and NP can act as both source and vector for various pollutants. As a result, contaminants attached to plastics may be transferred along the trophic levels as well (Chua et al. 2014; Wardrop et al. 2016; Tanaka et al. 2015; Gaylor et al. 2013).

The sediment-water interface is an important aspect in ecotoxicity of MP and NP with other pollutants. It is the site where gradients in physical, chemical and biological properties are the greatest (Santschi et al. 1990). As such, the distribution between sediments and water affects the transport of various pollutants such as POPs and heavy metals. This process is influenced by various factors such as salinity, presence of colloids, redox potential and pH (Means 1995; Baker et al. 1986; Eggleton and Thomas 2004). Due to weathering processes and formation of biofilm on the surface of MP and NP, the sedimentation rate of MP and NP is higher, resulting in higher concentration of the particles in sediments compared to water column. Larger MP with densities greater than water generally remain in the sediment, though they can be resuspended during periods of high water flow or other meteorological changes as well as bioturbation and human activities (Nizzetto et al. 2016; Eggleton and Thomas 2004). This leads to changes in the chemical properties and habitat conditions in the sediment, and thus, the adsorption and release of pollutants from MP and NP (Kleinteich et al. 2018). In addition, MP and NP are known to have higher adsorption and desorption of organic pollutants compared to natural sediments (Wang and Wang 2018). This has potential adverse implications on aquatic organisms, particularly sediment feeders as they might mistake MP and NP as food (Eerkes-Medrano et al. 2015).

Ma et al. (2016) assessed the effects of MP and NP on toxicity, bioaccumulation and environmental fate of phenanthrene in freshwaters and reported that the combined toxicity of 50 nm NP and phenanthrene to *Daphnia magna* showed an additive effect. During a 14-day incubation period, the presence of NP significantly increased bioaccumulation of phenanthrene-derived residues in the daphnid body and inhibited the removal and transformation of the chemical in the medium, whereas 10 mm MP showed negligible adverse effects on the daphnid. Besseling et al. (2012) showed that PS at low concentration (0.074% DW) enhanced bioaccumulation of PCBs in lugworms (*Arenicola marina*) by a factor of 1.1–3.6, in which PS had statistically significant effects on the fitness of lugworms and bioaccumulation.

Paul-Pont et al. (2016) assessed the combined toxicity of MP and fluoranthene in mussels, where histopathological damages and increased levels of antioxidant markers were recorded. Haghi and Banaee (2017) reported an increase in aspartate aminotransferase (AST), alkaline phosphatase (ALP), creatine phosphokinase (CPK), alanine aminotransferase (ALT), lactate dehydrogenase (LDH) and glucose levels with a decrease in total protein, globulin, cholesterol and triglyceride levels and gamma glutamyl transferase activity in common carp (*Cyprinus carpio*) exposed to a mixture of paraquat and MP. In another study on African catfish (*Clarias gariepinus*), MP was found to influence the effects of phenanthrene on the degree of tissue change (DTC) in the gill; plasma concentrations of cholesterol, high-density lipoprotein (HDL), total protein (TP), albumin and globulin; and the transcription levels of *fushi* tarazu-factor 1 (*ftz-f1*), gonadotropin-releasing hormone (GnRH), 11 β-hydroxysteroid dehydrogenase type 2 (*11β-hsd2*) and liver glycogen stores (Karami et al. 2016). However, in a study using common goby (*Pomatoschistus microps*), MP delayed pyrene-induced fish mortality and elevated the concentration of bile pyrene metabolites. Concurrent exposure to both MP and pyrene did not increase significantly the inhibitory effect on acetylcholinesterase (AChE) in the fish, indicating that the mechanism for AChE inhibition appeared to be different for pyrene and MP (Oliveira et al. 2013).

Sinche (2010) reported that the toxicity of phenol in the brine shrimp *Artemia* sp., exposed for 48 and 72 h, was reduced by the addition of 3 μm PS spheres to phenol solutions. The author suggested that adsorption of phenol to PS beads is supported by other plastic congener profiles, protecting the brine shrimp against toxic levels of phenol. Another study documented that POP concentrations in the plastic did not differ significantly between the high and medium plastic ingestion groups in a study using northern fulmars (*Fulmarus glacialis*). It was suggested that MP is more likely to act as a passive sampler, in which POPs transferred from bird lipids to MP, rather than as a vector of POP due to higher fugacity of POPs in biota lipids compared to MP (Herzke et al. 2016).

There have been several reported studies on the co-toxicity of MP and heavy metals. For instance, Akhbarizadeh et al. (2018) reported a positive relationship between MP and heavy metal contents in the fish *Alepes djedaba* and *Platycephalus indicus* but a negative relationship in *Epinephelus coioides*. The authors suggested that the accumulation of MPs and heavy metals in different fish species depends on various factors such as type of habitat, feeding strategy, physiological characteristics of an organism as well as the physiochemical behaviour and characteristics of pollutant.

Kim et al. (2017) found that nickel-fixed carboxyl-functionalized PS (PS-COOH) induced higher rate of immobilization of *Daphnia magna* compared to nickel (Ni)-fixed PS and Ni alone acute toxicity tests at all the MP concentrations tested. The 48 h EC_{50} values for PS and PS-COOH were 42.78 and 25.96 mg/mL, respectively, compared to 48 h EC_{50} value for nickel at 3.85 mg/L. The authors postulated that the functional groups of MP may influence the toxicity of heavy metal-fixed MP, and thus, the toxicity trends may vary even for the same pollutant.

Significant inhibition of brain AChE activity (64–76%) and significant increase of lipid peroxidation levels in the brain (2.9–3.4 folds) and muscle (2.2–2.9 folds) were reported in European seabass *Dicentrarchus labrax* exposed to a mixture of inorganic mercury (Hg) and MP (Barboza et al. 2017). However, a concentration-dependent accumulation of Hg in the brain and muscle of the fish was not observed. The same study also found that the decay of Hg in water was correlated to increasing MP concentrations.

Methylmercury, an organic mercury (Hg), in combination with several types of POPs with MP induced a much higher level of vacuolization in liver tissues of zebrafish (*Danio rerio*) compared to fish fed with just Hg- and POP-contaminated feed, but no vacuolization was observed in fish fed with just MP (Rainieri et al. 2018). Besides that, fish fed with feeds contaminated with a mixture of Hg, POPs and MP as well as just Hg and POPs demonstrated an overexpression of all the tested genes in the liver and brain, whereas MP alone did not induce any significant effects on gene expression. The results demonstrated that the toxicity of combination of MP and pollutants could be greater than that pollutants alone.

There has been at least one reported study on the combined effect of MP and NP, and other pollutants on microalgae. Davarpanah and Guilhermino (2015) assessed the toxic effects of MP in combination with copper (Cu) on the marine microalga *Tetraselmis chui*. However, no significant effect of MP on Cu toxicity in the microalga in terms of average specific growth rate was shown. Clearly, there is a need for further investigations on the impact of MP and NP on the toxicity of heavy metals in microalgae (see Sect. 5).

5 Future Research Directions

Microalgae play an important role as primary producers of food chain. Besides that, they have many other uses such as food additives, animal feeds, potential medicinal properties as well as in bioremediation of contaminated environment (Chu 2012; Chu and Phang 2016; Phang et al. 2015). Yet there have been limited studies on the potential impacts of MP and NP, particularly co-toxicity with other pollutants, on microalgae.

One priority co-pollutant that should be given attention in further studies on the potential toxic effects of MP and NP on microalgae is heavy metals. Heavy metals (e.g. copper and zinc) are known to cause a wide range of toxic effects on microalgae, including induction of oxidative stress (Hamed et al. 2017), growth inhibition (Wan et al. 2018), biochemical and metabolomic disturbances (Gonçalves et al. 2018) and decreased photosynthetic activity (Couet et al. 2018). However, metal adsorption and desorption in relation to MP and NP, and how this may influence metal toxicity in microalgae, has not been studied. There has been evidence on the adsorption of heavy metals such as Ni, cadmium (Cd), lead (Pb), Cu, zinc (Zn) and titanium (Ti) to MP and NP from surface sediments, as reported in a study on Beijiang River littoral zone (Wang et al. 2017b). There was another study that showed significant positive correlation between heavy metals, such as vanadium (V), Cu, chromium (Cr), Ni and Pb, and MP in coastal sediments of Kharg Island, indicating possible adsorption of heavy metals by the MP present (Akhbarizadeh et al. 2017). Field studies should be conducted to correlate heavy metal and MP contents in the water column and sediments with the abundance of phytoplankton. Laboratory studies focussing on the distribution of dissolved metals and the fraction adsorbed to MP as well as metal uptake by algal cells are required in attempts to elucidate the mechanistic aspects of combined toxicity of MP and heavy metals in microalgae.

Another potential area of further research is concerning the co-toxicity of MP and NP with other particulate pollutants, especially metal oxide nanoparticles. Examples of metal oxide nanoparticles include copper oxide nanoparticles (CuO NP), silver nanoparticles (Ag NP), zinc oxide nanoparticles (ZnO NP) and titanium dioxide nanoparticles (TiO_2 NP). Metal nanoparticles have been shown to cause a wide range of adverse effects on microalgae. For instance, CuO NP have been shown to have growth inhibitory effect on a tropical *Chlorella*, although the effect was less severe than dissolved copper ions (Wan et al. 2018). The CuO NP was also found to penetrate *Chlorella* sp. through the cell. In another study, Oukarroum et al. (2012) reported that Ag NP caused a decrease in chlorophyll content, viable algal cells, increased ROS formation and lipid peroxidation in

Chlorella vulgaris and *Dunaliella tertiolecta*. The combined effects of metal oxide nanoparticles and MP and NP on microalgae have not been reported. However, there has been a recent study on the synergistic toxic effect of gold nanoparticles and MP on *Daphnia magna*, in causing mortality and reproduction impairment (Pacheco et al. 2018).

There has been hardly any information on the leaching of chemical additives, especially plasticizers from MP and NP. No studies have been carried out so far on microalgae to assess the toxic effects of plasticizers that leach directly from MP and NP. However, there have been several reported studies on the toxic effects of plasticizers added directly into the algal culture medium. For example, individual and combined treatments of BPA and DEHP were shown to be toxic to the marine toxic dinoflagellate *Alexandrium pacificum,* adversely affecting its photosynthetic activity (M'Rabet et al. 2018). Exposure to DBP was also found to cause large cell vacuolization, detachment of plasma membrane from cell wall with distortion to the shape of plasma membrane as well as destruction of various organelles such as chloroplast and protein rings in *Scenedesmus obliquus* and *Chlorella pyrenoidosa* (Gu et al. 2017). In assessing possible toxic effects of plasticizers that leach from MP and NP, long-term exposure of the microalgae to the toxicants will be required. This is to allow monitoring of the amounts of the chemicals that leach from MP and NP, which may take a long time, in causing any toxic effect on microalgae.

The toxicity of MP and NP is an emerging field that has attracted much interest amongst researchers, as reflected by the vast amounts of literature in recent years. However, studies related to microalgae are still limited, despite their key role as primary producers in the ecosystem. The proposed areas for further research highlighted above should be pursued to gain insights into the potential impacts of MP and NP on microalgae.

6   Conclusion

Plastics in the environment have the potential to fragment into MP and NP. These microscopic particles can cause various adverse effects on a broad spectrum of organisms. However, there have been very few studies that focused on the effects of MP and NP on microalgae. The standardization of definition of MP and NP as well as quantification of abundances of the particles in the environment is essential to enable easier and faster communication of research data. Evidence of increased toxicity with the concurrent presence of MP and NP with various chemicals such as POPs and heavy metals has also been documented in several organisms. However, there is a scarcity of research done to investigate such combined effects on microalgae, which are the primary producers of food chain and an important group of organisms with many potential commercial applications.

7 Summary

Plastics are able to degrade to micro- and nanosize particles known as microplastics (MP) and nanoplastics (NP), respectively. These minute particles have been shown to cause various adverse effects on aquatic organisms. This review covers the distribution of MP and NP in aquatic ecosystems, with emphasis on their effects on microalgae as well as co-toxicity of MP and NP with other pollutants. Some potential areas for future research are proposed to address the knowledge gaps in this field, especially with regard to the combined toxic effects of MP and NP with other pollutants on microalgae.

Acknowledgements The authors would like to acknowledge the internal grant (Grant No. IMU 377/2017) from the International Medical University in supporting the research project on the effects of MP and NP on microalgae.

References

Acosta-Coley I, Olivero-Verbel J (2015) Microplastic resin pellets on an urban tropical beach in Colombia. Environ Monit Assess 187(7):435

Akhbarizadeh R, Moore F, Keshavarzi B, Moeinpour A (2017) Microplastics and potentially toxic elements in coastal sediments of Iran's main oil terminal (Khark Island). Environ Pollut 220:720–731

Akhbarizadeh R, Moore F, Keshavarzi B (2018) Investigating a probable relationship between microplastics and potentially toxic elements in fish muscles from northeast of Persian Gulf. Environ Pollut 232:154–163

Andrady AL (2011) Microplastics in the marine environment. Mar Pollut Bull 62(8):1596–1605

Baker JE, Capel PD, Eisenreich SJ (1986) Influence of colloids on sediment-water partition coefficients of polychlorobiphenyl congeners in natural waters. Environ Sci Technol 20 (11):1136–1143

Barboza LGA, Vieira LR, Branco V, Figueiredo N, Carvalho F, Carvalho C, Guilhermino L (2017) Microplastics cause neurotoxicity, oxidative damage and energy-related changes and interact with the bioaccumulation of mercury in the European seabass, *Dicentrarchus labrax* (Linnaeus, 1758). Aquat Toxicol 195:49–57

Bergami E, Bocci E, Vannuccini ML, Monopoli M, Salvati A, Dawson KA, Corsi I (2016) Nano-sized polystyrene affects feeding, behavior and physiology of brine shrimp *Artemia franciscana* larvae. Ecotoxicol Environ Saf 123:18–25

Besseling E, Wegner A, Foekema EM, Van Den Heuvel-Greve MJ, Koelmans AA (2012) Effects of microplastic on fitness and PCB bioaccumulation by the lugworm *Arenicola marina* (L.). Environ Sci Technol 47(1):593–600

Besseling E, Wang B, Lürling M, Koelmans AA (2014) Nanoplastic affects growth of *S. obliquus* and reproduction of *D. magna*. Environ Sci Technol 48(20):12336–12343

Bhattacharya P, Lin S, Turner JP, Ke PC (2010) Physical adsorption of charged plastic nanoparticles affects algal photosynthesis. J Phys Chem C 114(39):16556–16561

Booth AM, Hansen BH, Frenzel M, Johnsen H, Altin D (2015) Uptake and toxicity of methylmethacrylate-based nanoplastic particles in aquatic organisms. Environ Toxicol Chem 35(7):1641–1649

Browne MA, Galloway T, Thompson R (2007) Microplastic – an emerging contaminant of potential concern? Integr Environ Assess Manag 3:559–561

Browne MA, Dissanayake A, Galloway TS, Lowe DM, Thompson RC (2008) Ingested microscopic plastic translocates to the circulatory system of the mussel, *Mytilus edulis* (L.). Environ Sci Technol 42(13):5026–5031

Browne MA, Galloway TS, Thompson RC (2010) Spatial patterns of plastic debris along estuarine shorelines. Environ Sci Technol 44:3404–3409

Casado MP, Macken A, Byrne HJ (2013) Ecotoxicological assessment of silica and polystyrene nanoparticles assessed by a multitrophic test battery. Environ Int 51:97–105

Castañeda RA, Avlijas S, Simard MA, Ricciardi A (2014) Microplastic pollution in St. Lawrence River sediments. Can J Fish Aquat Sci 71(12):1767–1771

Chae Y, Kim D, Kim SW, An YJ (2018) Trophic transfer and individual impact of nano-sized polystyrene in a four-species freshwater food chain. Sci Rep 8(1):284

Chen Q, Gundlach M, Yang S, Jiang J, Velki M, Yin D, Hollert H (2017) Quantitative investigation of the mechanisms of microplastics and nanoplastics toward zebrafish larvae locomotor activity. Sci Total Environ 584:1022–1031

Chu WL (2012) Biotechnological applications of microalgae. IeJSME 6(1):S24–S37

Chu WL, Phang SM (2016) Marine algae as a potential source for anti-obesity agents. Mar Drugs 14 (12):222

Chua EM, Shimeta J, Nugegoda D, Morrison PD, Clarke BO (2014) Assimilation of polybrominated diphenyl ethers from microplastics by the marine amphipod, *Allorchestes compressa*. Environ Sci Technol 48(14):8127–8134

Cincinelli A, Scopetani C, Chelazzi D, Lombardini E, Martellini T, Katsoyiannis A, Corsolini S (2017) Microplastic in the surface waters of the Ross Sea (Antarctica): occurrence, distribution and characterization by FTIR. Chemosphere 175:391–400

Claessens M, Meester SD, Landuyt LV, Clerck KD, Janssen CR (2011) Occurrence and distribution of microplastics in marine sediments along the Belgian coast. Mar Pollut Bull 62 (10):2199–2204

Cole M, Lindeque P, Halsband C, Galloway TS (2011) Microplastics as contaminants in the marine environment: a review. Mar Pollut Bull 62(12):2588–2597

Cole M, Lindeque P, Fileman E, Halsband C, Goodhead R, Moger J, Galloway TS (2013) Microplastic ingestion by zooplankton. Environ Sci Technol 47(12):6646–6655

Cole M, Lindeque P, Fileman E, Halsband C, Galloway TS (2015) The impact of polystyrene microplastics on feeding, function and fecundity in the marine copepod *Calanus helgolandicus*. Environ Sci Technol 49(2):1130–1137

Collignon A, Hecq JH, Galgani F, Collard F, Goffart A (2014) Annual variation in neustonic micro- and meso-plastic particles and zooplankton in the Bay of Calvi (Mediterranean–Corsica). Mar Pollut Bull 79(1):293–298

Costa MF, Do Sul JAI, Silva-Cavalcanti JS, Araújo MCB, Spengler Â, Tourinho PS (2010) On the importance of size of plastic fragments and pellets on the strandline: a snapshot of a Brazilian beach. Environ Monit Assess 168(1–4):299–304

Couet D, Pringault O, Bancon-Montigny C, Briant N, Poulichet FE, Delpoux S, Amzil Z (2018) Effects of copper and butyltin compounds on the growth, photosynthetic activity and toxin production of two HAB dinoflagellates: the planktonic *Alexandrium catenella* and the benthic *Ostreopsis* cf. *ovata*. Aquat Toxicol 196:154–167

Crawford CB, Quinn B (2017) Plastic production, waste and legislation. Microplastic pollutants. Elsevier, Amsterdam, pp 39–56

da Costa JP, Santos PS, Duarte AC, Rocha-Santos T (2016) (Nano)plastics in the environment – sources, fates and effects. Sci Total Environ 566:15–26

Davarpanah E, Guilhermino L (2015) Single and combined effects of microplastics and copper on the population growth of the marine microalgae *Tetraselmis chui*. Estuar Coast Shelf Sci 167:269–275

De Troyer N (2015) Occurrence and distribution of microplastics in the Scheldt River. Universiteit Gent, Ghent

Dedman CJ (2014) Investigating microplastic ingestion by zooplankton. University of Exeter, Exeter

Della Torre C, Bergami E, Salvati A, Faleri C, Cirino P, Dawson KA, Corsi I (2014) Accumulation and embryotoxicity of polystyrene nanoparticles at early stage of development of sea urchin embryos *Paracentrotus lividus*. Environ Sci Technol 48(20):12302–12311

Eerkes-Medrano D, Thompson RC, Aldridge DC (2015) Microplastics in freshwater systems: a review of the emerging threats, identification of knowledge gaps and prioritisation of research needs. Water Res 75:63–82

Eggleton J, Thomas KV (2004) A review of factors affecting the release and bioavailability of contaminants during sediment disturbance events. Environ Int 30(7):973–980

Esiukova E (2017) Plastic pollution on the Baltic beaches of Kaliningrad region, Russia. Mar Pollut Bull 114(2):1072–1080

Fok L, Cheung PK, Tang G, Li WC (2017) Size distribution of stranded small plastic debris on the coast of Guangdong, South China. Environ Pollut 220:407–412

Free CM, Jensen OP, Mason SA, Eriksen M, Williamson NJ, Boldgiv B (2014) High-levels of microplastic pollution in a large, remote, mountain lake. Mar Pollut Bull 85(1):156–163

Gaylor MO, Harvey E, Hale RC (2013) Polybrominated diphenyl ether (PBDE) accumulation by earthworms (*Eisenia fetida*) exposed to biosolids-, polyurethane foam microparticle-, and penta-BDE-amended soils. Environ Sci Technol 47(23):13831–13839

Gonçalves S, Kahlert M, Almeida SF, Figueira E (2018) Assessing Cu impacts on freshwater diatoms: biochemical and metabolomic responses of *Tabellaria flocculosa* (Roth) Kützing. Sci Total Environ 625:1234–1246

Gosden E (2016) More plastic than fish in the oceans by 2050, report warns. The Telegraph. http://www.telegraph.co.uk/news/earth/environment/12108522/More-plastic-than-fish-in-the-oceans-by-2050-report-warns.html. Accessed 26 Feb 2018

Graca B, Szewc K, Zakrzewska D, Dołęga A, Szczerbowska-Boruchowska M (2017) Sources and fate of microplastics in marine and beach sediments of the Southern Baltic Sea – a preliminary study. Environ Sci Pollut Res 24(8):7650–7661

Graham ER, Thompson JT (2009) Deposit- and suspension-feeding sea cucumbers (*Echinodermata*) ingest plastic fragments. J Exp Mar Biol Ecol 368:22–29

Green DS (2016) Effects of microplastics on European flat oysters, *Ostrea edulis* and their associated benthic communities. Environ Pollut 216:95–103

Green DS, Boots B, Sigwart J, Jiang S, Rocha C (2016) Effects of conventional and biodegradable microplastics on a marine ecosystem engineer (*Arenicola marina*) and sediment nutrient cycling. Environ Pollut 208:426–434

Greven AC, Merk T, Karagöz F, Mohr K, Klapper M, Jovanović B, Palić D (2016) Polycarbonate and polystyrene nanoplastic particles act as stressors to the innate immune system of fathead minnow (*Pimephales promelas*). Environ Toxicol Chem 35(12):3093–3100

Gu S, Zheng H, Xu Q, Sun C, Shi M, Wang Z, Li F (2017) Comparative toxicity of the plasticizer dibutyl phthalate to two freshwater algae. Aquat Toxicol 191:122–130

Gutow L, Eckerlebe A, Giménez L, Saborowski R (2015) Experimental evaluation of seaweeds as a vector for microplastics into marine food webs. Environ Sci Technol 50(2):915–923

Haghi BN, Banaee M (2017) Effects of micro-plastic particles on paraquat toxicity to common carp (*Cyprinus carpio*): biochemical changes. Int J Environ Sci Technol 14(3):521–530

Hamed SM, Zinta G, Klöck G, Asard H, Selim S, Abd Elgawad H (2017) Zinc-induced differential oxidative stress and antioxidant responses in *Chlorella sorokiniana* and *Scenedesmus acuminatus*. Ecotoxicol Environ Saf 140:256–263

Hammer J, Kraak MH, Parsons JR (2012) Plastics in the marine environment: the dark side of a modern gift. Rev Environ Contam Toxicol 220:1–44

Hartline NL, Bruce NJ, Karba SN, Ruff EO, Sonar SU, Holden PA (2016) Microfiber masses recovered from conventional machine washing of new or aged garments. Environ Sci Technol 50(21):11532–11538

Haseler M, Schernewski G, Balciunas A, Sabaliauskaite V (2017) Monitoring methods for large micro-and meso-litter and applications at Baltic beaches. J Coast Conserv 22(1):27–50

Hentschel LH (2015) Understanding species-microplastics interactions: a laboratory study on the effects of microplastics on the Azorean barnacle, Megabalanus azoricus. University of Akureyri, Akureyri

Herzke D, Anker-Nilssen T, Nøst TH, Götsch A, Christensen-Dalsgaard S, Langset M, Koelmans AA (2016) Negligible impact of ingested microplastics on tissue concentrations of persistent organic pollutants in northern fulmars off coastal Norway. Environ Sci Technol 50(4):1924–1933

Hidalgo-Ruz V, Gutow L, Thompson RC, Thiel M (2012) Microplastics in the marine environment: a review of the methods used for identification and quantification. Environ Sci Technol 46 (6):3060–3075

Horton AA, Svendsen C, Williams RJ, Spurgeon DJ, Lahive E (2017a) Large microplastic particles in sediments of tributaries of the River Tames, UK–Abundance, sources and methods for effective quantification. Mar Pollut Bull 114(1):218–226

Horton AA, Walton A, Spurgeon DJ, Lahive E, Svendsen C (2017b) Microplastics in freshwater and terrestrial environments: evaluating the current understanding to identify the knowledge gaps and future research priorities. Sci Total Environ 586:127–141

Isobe A, Uchiyama-Matsumoto K, Uchida K, Tokai T (2017) Microplastics in the Southern Ocean. Mar Pollut Bull 114(1):623–626

Jemec A, Horvat P, Kunej U, Bele M, Kržan A (2016) Uptake and effects of microplastic textile fibers on freshwater crustacean Daphnia magna. Environ Pollut 219:201–209

Jeong CB, Won EJ, Kang HM, Lee MC, Hwang DS, Hwang UK, Lee JS (2016) Microplastic size-dependent toxicity, oxidative stress induction, and p-JNK and p-p38 activation in the monogonont rotifer (Brachionus koreanus). Environ Sci Technol 50(16):8849–8857

Jeong CB, Kang HM, Lee MC, Kim DH, Han J, Hwang DS, Lee JS (2017) Adverse effects of microplastics and oxidative stress-induced MAPK/Nrf2 pathway-mediated defense mechanisms in the marine copepod Paracyclopina nana. Sci Rep 7:41323

Kaplan S (2016) By 2050, there will be more plastic than fish in the world's oceans, study says. The Washington Post. https://www.washingtonpost.com/news/morning-mix/wp/2016/01/20/by-2050-there-will-be-more-plastic-than-fish-in-the-worlds-oceans-study-says/?utm_term=. 0f6c653beaab. Accessed 26 Feb 2018

Karami A, Romano N, Galloway T, Hamzah H (2016) Virgin microplastics cause toxicity and modulate the impacts of phenanthrene on biomarker responses in African catfish (Clarias gariepinus). Environ Res 151:58–70

Katzenberger TD (2015) Assessing the biological effects of exposure to microplastics in the three-spined stickleback (Gasterosteus aculeatus) (Linnaeus 1758). University of York, York

Kim D, Chae Y, An YJ (2017) Mixture toxicity of nickel and microplastics with different functional groups on Daphnia magna. Environ Sci Technol 51(21):12852–12858

Klein S, Worch E, Knepper TP (2015) Occurrence and spatial distribution of microplastics in river shore sediments of the Rhine-main area in Germany. Environ Sci Technol 49(10):6070–6076

Kleinteich J, Seidensticker S, Marggrander N, Zarfl C (2018) Microplastics reduce short-term effects of environmental contaminants. Part II: polyethylene particles decrease the effect of polycyclic aromatic hydrocarbons on microorganisms. Int J Environ Res Publ Health 15(2):287

Lagarde F, Olivier O, Zanella M, Daniel P, Hiard S, Caruso A (2016) Microplastic interactions with freshwater microalgae: hetero-aggregation and changes in plastic density appear strongly dependent on polymer type. Environ Pollut 215:331–339

Lechner A, Keckeis H, Lumesberger-Loisl F, Zens B, Krusch R, Tritthart M, Schludermann E (2014) The Danube so colourful: a potpourri of plastic litter outnumbers fish larvae in Europe's second largest river. Environ Pollut 188:177–181

Lee KW, Shim WJ, Kwon OY, Kang JH (2013) Size-dependent effects of micro polystyrene particles in the marine copepod Tigriopus japonicus. Environ Sci Technol 47(19):11278–11283

Lithner D (2011) Environmental and health hazards of chemicals in plastic polymers and products. University of Gothenburg, Gothenburg

Long M, Paul-Pont I, Hégaret H, Moriceau B, Lambert C, Huvet A, Soudant P (2017) Interactions between polystyrene microplastics and marine phytoplankton lead to species-specific hetero-aggregation. Environ Pollut 228:454–463

Lu Y, Zhang Y, Deng Y, Jiang W, Zhao Y, Geng J, Ren H (2016) Uptake and accumulation of polystyrene microplastics in zebrafish (*Danio rerio*) and toxic effects in liver. Environ Sci Technol 50(7):4054–4060

Lusher AL, Tirelli V, O'Connor I, Officer R (2015) Microplastics in Arctic polar waters: the first reported values of particles in surface and sub-surface samples. Sci Rep 5:14947

M'Rabet C, Pringault O, Zmerli-Triki H, Gharbia HB, Couet D, Yahia OKD (2018) Impact of two plastic-derived chemicals, the Bisphenol A and the di-2-ethylhexyl phthalate, exposure on the marine toxic dinoflagellate *Alexandrium pacificum*. Mar Pollut Bull 126:241–249

Ma Y, Huang A, Cao S, Sun F, Wang L, Guo H, Ji R (2016) Effects of nanoplastics and microplastics on toxicity, bioaccumulation, and environmental fate of phenanthrene in fresh water. Environ Pollut 219:166–173

Mahon AM, O'Connell B, Healy MG, O'Connor I, Officer R, Nash R, Morrison L (2016) Microplastics in sewage sludge: effects of treatment. Environ Sci Technol 51(2):810–818

Mani T, Hauk A, Walter U, Burkhardt-Holm P (2014) Microplastics profile along the Rhine River. Sci Rep 5:17988–17988

Mato Y, Isobe T, Takada H, Kanehiro H, Ohtake C, Kaminuma T (2001) Plastic resin pellets as a transport medium for toxic chemicals in the marine environment. Environ Sci Technol 35 (2):318–324

Mattsson K, Ekvall MT, Hansson LA, Linse S, Malmendal A, Cedervall T (2014) Altered behavior, physiology, and metabolism in fish exposed to polystyrene nanoparticles. Environ Sci Technol 49(1):553–561

McCormick AR, Hoellein TJ, London MG, Hittie J, Scott JW, Kelly JJ (2016) Microplastic in surface waters of urban rivers: concentration, sources, and associated bacterial assemblages. Ecosphere 7(11):e01556

Means JC (1995) Influence of salinity upon sediment-water partitioning of aromatic hydrocarbons. Mar Chem 51(1):3–16

Miyake H, Shibata H, Furushima Y (2011) Deep-sea litter study using deep-sea observation tools. In: Interdisciplinary studies on environmental chemistry-marine environmental modeling and analysis. Terrapub, Tokyo, pp 261–269

Morét-Ferguson S, Law KL, Proskurowski G, Murphy EK, Peacock EE, Reddy CM (2010) The size, mass, and composition of plastic debris in the western North Atlantic Ocean. Mar Pollut Bull 60(10):1873–1878

Nizzetto L, Bussi G, Futter MN, Butterfield D, Whitehead PG (2016) A theoretical assessment of microplastic transport in river catchments and their retention by soils and river sediments. Environ Sci Process Impact 18(8):1050–1059

Nolte TM, Hartmann NB, Kleijn JM, Garnæs J, van de Meent D, Hendriks AJ, Baun A (2017) The toxicity of plastic nanoparticles to green algae as influenced by surface modification, medium hardness and cellular adsorption. Aquat Toxicol 183:11–20

Oliveira M, Ribeiro A, Hylland K, Guilhermino L (2013) Single and combined effects of microplastics and pyrene on juveniles (0+ group) of the common goby *Pomatoschistus microps* (Teleostei, gobiidae). Ecol Indic 34:641–647

Oukarroum A, Bras S, Perreault F, Popovic R (2012) Inhibitory effects of silver nanoparticles in two green algae, *Chlorella vulgaris* and *Dunaliella tertiolecta*. Ecotoxicol Environ Saf 78:80–85

Pacheco A, Martins A, Guilhermino L (2018) Toxicological interactions induced by chronic exposure to gold nanoparticles and microplastics mixtures in *Daphnia magna*. Sci Total Environ 628–629:474–483

Paul-Pont I, Lacroix C, Fernández CG, Hégaret H, Lambert C, Le Goïc N, Guyomarch J (2016) Exposure of marine mussels *Mytilus* spp. to polystyrene microplastics: toxicity and influence on fluoranthene bioaccumulation. Environ Pollut 216:724–737

Phang SM, Chu WL, Rabiei R (2015) Phycoremediation. In: Sahoo D, Seckbach J (eds) The algae world. Springer, Berlin, pp 357–389

Phillips C (2017) Ghostly encounters: dealing with ghost gear in the Gulf of Carpentaria. Geoforum 78:33–42

Rainieri S, Conlledo N, Larsen BK, Granby K, Barranco A (2018) Combined effects of microplastics and chemical contaminants on the organ toxicity of zebrafish (*Danio rerio*). Environ Res 162:135–143

Rehse S, Kloas W, Zarfl C (2016) Short-term exposure with high concentrations of pristine microplastic particles leads to immobilisation of *Daphnia magna*. Chemosphere 153:91–99

Retama I, Jonathan MP, Shruti VC, Veluman S, Sarkar SK, Roy PD, Rodríguez-Espinosa PF (2016) Microplastics in tourist beaches of Huatulco Bay, Pacific Coast of Southern Mexico. Mar Pollut Bull 113(1):530–535

Rosenkranz P, Chaudhry Q, Stone V, Fernandes TF (2009) A comparison of nanoparticle and fine particle uptake by *Daphnia magna*. Environ Toxicol Chem 28(10):2142–2149

Sadri SS, Thompson RC (2014) On the quantity and composition of floating plastic debris entering and leaving the Tamar Estuary, Southwest England. Mar Pollut Bull 1(1):55–60

Santschi P, Höhener P, Benoit G, Buchholtz-ten Brink M (1990) Chemical processes at the sediment-water interface. Mar Chem 30:269–315

Setälä O, Norkko J, Lehtiniemi M (2016) Feeding type affects microplastic ingestion in a coastal invertebrate community. Mar Pollut Bull 102(1):95–101

Sinche F (2010) Impact of microparticle concentration levels upon toxicity of phenol to *Artemia*. Clemson University, Clemson

Sjollema SB, Redondo-Hasselerharm P, Leslie HA, Kraak MH, Vethaak AD (2016) Do plastic particles affect microalgal photosynthesis and growth? Aquat Toxicol 170:259–261

Sruthy S, Ramasamy EV (2017) Microplastic pollution in Vembanad Lake, Kerala, India: the first report of microplastics in lake and estuarine sediments in India. Environ Pollut 222:315–322

Su L, Xue Y, Li L, Yang D, Kolandhasamy P, Li D, Shi H (2016) Microplastics in Taihu Lake, China. Environ Pollut 216:711–719

Sussarellu R, Suquet M, Thomas Y, Lambert C, Fabioux C, Pernet MEJ, Corporeau C (2016) Oyster reproduction is affected by exposure to polystyrene microplastics. Proc Natl Acad Sci U S A 113(9):2430–2435

Tagg AS, Sapp M, Harrison JP, Ojeda JJ (2015) Identification and quantification of microplastics in wastewater using focal plane array-based reflectance micro-FT-IR imaging. Anal Chem 87 (12):6032–6040

Talley KJ (2015) The effect of microplastic fibers on the freshwater amphipod, Hyalella azteca. Clemson University, Clemson

Talsness CE, Andrade AJ, Kuriyama SN, Taylor JA, vom Saal FS (2009) Components of plastic: experimental studies in animals and relevance for human health. Philos Trans R Soc Lond B Biol Sci 364(1526):2079–2096

Tanaka K, Takada H, Yamashita R, Mizukawa K, Fukuwaka MA, Watanuki Y (2015) Facilitated leaching of additive-derived PBDEs from plastic by seabirds' stomach oil and accumulation in tissues. Environ Sci Technol 49(19):11799–11807

Tosetto L (2015) Impacts of microplastics on coastal biota and the potential for trophic transfer. Macquarie University, Sydney

Tsang YY, Mak CW, Liebich C, Lam SW, Sze ET, Chan KM (2017) Microplastic pollution in the marine waters and sediments of Hong Kong. Mar Pollut Bull 115(1–2):20–28

Urgert W (2015) Microplastics in the rivers Meuse and Rhine. Open University of the Netherlands, Heerlen

Van Cauwenberghe L, Devriese L, Galgani F, Robbens J, Janssen CR (2015) Microplastics in sediments: a review of techniques, occurrence and effects. Mar Environ Res 111:5–17

Vianello A, Boldrin A, Guerriero P, Moschino V, Rella R, Sturaro A, Da Ros L (2013) Microplastic particles in sediments of lagoon of Venice, Italy: first observations on occurrence, spatial patterns and identification. Estuar Coast Shelf Sci 130:54–61

Visne A, Bat L (2016) Plastic pollution in Sinop Sarikum lagoon coast in the Southern Black Sea [Abstract]. Rapp Comm Int Mer Médit 41

Von Moos N, Burkhardt-Holm P, Köhler A (2012) Uptake and effects of microplastics on cells and tissue of the blue mussel *Mytilus edulis* L. after an experimental exposure. Environ Sci Technol 46(20):11327–11335

Wagner M, Scherer C, Alvarez-Muñoz D, Brennholt N, Bourrain X, Buchinger S, Rodriguez-Mozaz S (2014) Microplastics in freshwater ecosystems: what we know and what we need to know. Environ Sci Eur 26(1):12

Waller CL, Griffiths HJ, Waluda CM, Thorpe SE, Loaiza I, Moreno B, Hughes KA (2017) Microplastics in the Antarctic marine system: an emerging area of research. Sci Total Environ 598:220–227

Wan JK, Chu WL, Kok YY, Cheong KW (2018) Assessing the toxicity of copper oxide nanoparticles and copper sulfate in a tropical *Chlorella*. J Appl Phycol. https://doi.org/10.1007/s10811-018-1408-3

Wang W, Wang J (2018) Different partition of polycyclic aromatic hydrocarbon on environmental particulates in freshwater: microplastics in comparison to natural sediment. Ecotoxicol Environ Saf 147:648–655

Wang W, Ndungu AW, Li Z, Wang J (2017a) Microplastics pollution in inland freshwaters of China: a case study in urban surface waters of Wuhan, China. Sci Total Environ 575:369–1374

Wang J, Peng J, Tan Z, Gao Y, Zhan Z, Chen Q, Cai L (2017b) Microplastics in the surface sediments from the Beijiang River littoral zone: composition, abundance, surface textures and interaction with heavy metals. Chemosphere 171:248–258

Wardrop P, Shimeta J, Nugegoda D, Morrison PD, Miranda A, Tang M, Clarke BO (2016) Chemical pollutants sorbed to ingested microbeads from personal care products accumulate in fish. Environ Sci Technol 50(7):4037–4044

Watts AJ, Urbina MA, Goodhead R, Moger J, Lewis C, Galloway TS (2016) Effect of microplastic on the gills of the shore crab *Carcinus maenas*. Environ Sci Technol 50(10):5364–5369

Wearden G (2016) More plastic than fish in the sea by 2050, says Ellen MacArthur. The Guardian. https://www.theguardian.com/business/2016/jan/19/more-plastic-than-fish-in-the-sea-by-2050-warns-ellen-macarthur. Accessed 26 Feb 2018

Wegner A, Besseling E, Foekema EM, Kamermans P, Koelmans AA (2012) Effects of nano-polystyrene on the feeding behavior of the blue mussel (*Mytilus edulis* L.). Environ Toxicol Chem 31(11):2490–2497

Wessel CC, Lockridge GR, Battiste D, Cebrian J (2016) Abundance and characteristics of microplastics in beach sediments: insights into microplastic accumulation in northern Gulf of Mexico estuaries. Mar Pollut Bull 109(1):178–183

Yokota K, Waterfield H, Hastings C, Davidson E, Kwietniewski E, Wells B (2017) Finding the missing piece of the aquatic plastic pollution puzzle: interaction between primary producers and microplastics. Limnol Oceanogr Lett 2(4):91–104

Yonkos LT, Friedel EA, Perez-Reyes AC, Ghosal S, Arthur CD (2014) Microplastics in four estuarine rivers in the Chesapeake Bay, USA. Environ Sci Technol 48(24):14195–14202

Young AM, Elliott JA (2016) Characterization of microplastic and mesoplastic debris in sediments from Kamilo Beach and Kahuku Beach, Hawai'i. Mar Pollut Bull 113(1):477–482

Zhang C, Chen X, Wang J, Tan L (2017) Toxic effects of microplastic on marine microalgae *Skeletonema costatum*: interactions between microplastic and algae. Environ Pollut 220:1282–1288

Index

Printed in the United States
By Bookmasters